Biotech's

DICTIONARY *Of* Forestry

Compiled & Edited by

Meena Francis

BIOTECH

2011

BIOTECH BOOKS

DELHI - 110035

Biotech's

DICTIONARY Of Forestry

Compiled & Edited by
Meena Francis

ISBN 81-7622-131-7

Published by : BIOTECH BOOKS
1123/74, Tri Nagar,
Delhi - 110 035
Phone: 27382765
e-mail: biotechbooks@yahoo.co.in

Showroom : 4762-63/23, Ansari Road, Darya Ganj,
New Delhi - 110 002
Phone: 23245578, 23244987

Printed at : Chawla Offset Printers
New Delhi - 110 052

PRINTED IN INDIA

PREFACE

Ten thousand years ago, about half of the surface of the earth was covered with trees. Now, that proportion is down to about one-third, or four billion hectares. What's worse, the rate of cutting has increased exponentially in recent years. Globally, we are now cutting down forests at the rate of 400 000 hectares a week. At this rate, it is estimated there will be no forests left by the year 2030.

But a forest is more than just the trees. A single forest is a complete, functioning ecosystem that supports innumerable plant and animal species as well as earth, water and air subsystems. The subsystems provide the essence of life of the forest and are in themselves a by-product of forest systems, all of which are reciprocal and interdependent. The further bad news is that 50 000 animal species disappear every year, many of them without ever having been classified or recorded. One of the main causes of this mass extinction is deforestation. Deforestation is also a major cause of soil erosion and water pollution, and the rampant destruction of the tropical rain forests in the southern hemisphere, according to some scientists, is contributing to the greenhouse effect, which in turn is leading to global warming. So why do we keep cutting the trees and what must we do to solve the problems created by deforestation? On the one hand, the answer is simple - reduce the number of trees being cut down on this planet. On the other hand, the answers are complex and involve the global, economic and political issues of wealth redistribution, poverty, land reform, overconsumption and overpopulation. Each of these is a complex political issue and each has a direct link to deforestation. None is easy to solve in the short term.

Most forestry experts agree, we must solve them and efforts are being made at both grassroots and international levels to do so. The aim is to improve our understanding of both the environment and the processes that underlie the Earth's support systems. Moreover, the impacts of human activity on natural environments are particularly interested. Therefore it can provide us a practicable solution to today's pressing environmental problems, so that a healthy, wealthy and sustainable environment can be enhanced and maintained.

- **abdomen**
  the third or posterior major division of the insect body; consists normally of nine or ten apparent segments; bears no functional legs in the adult stage, but may bear prolegs or false legs in the larval stage.

- **abiotic**
  1. living components of the environment.
  2. non-living components of the environment.

- **abiotic factors**
  the non-living components of the environment, such as air, rocks, soil, water, peat and plant litter.

- **abiotic pathogen**
  a nonliving, disease-causing entity, e.g., drought, salt, air pollutants etc.

- **aboral**
  relating to position: away from the mouth.

- **aboriginal resource unit**
  an investigated unit identified by the aboriginal communities/bands that provides resources for food or culture uses (e.g. ceremonies). Each site is described by its band, location and resource type, use and quality on a monthly basis. This information is confidential and not released without a band's permission.

- **abscission layer**
  layer of cells that forms at the base of each leaf petiole where it is attached to the twig when the veins that carry fluids into and out of the leaf gradually close off.

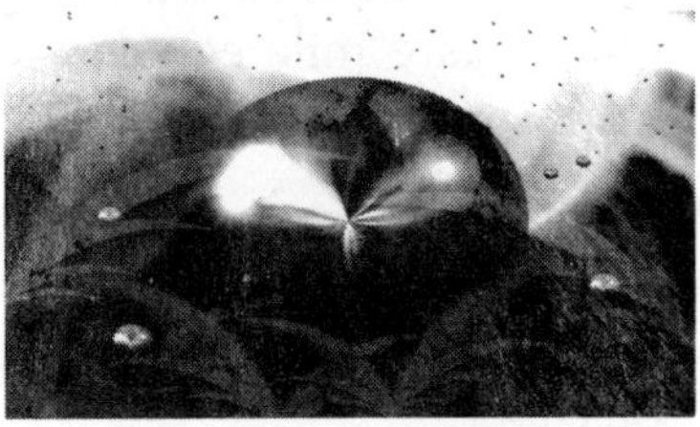

- **access management plan**
  an operational plan that shows how road construction, modification and deactivation will be carried out to protect or mitigate impacts on known resources or sensitive areas while maximising the efficacy of forest resource development.

- **acid rain**
  rainfall with a pH of less than 7.0. Long-term deposition of these acids is linked to adverse effects on aquatic organisms and plant life in areas with poor neutralising (buffering) capacity.

- **acre**
  1. area of land measuring about 43,560 square feet. A square 1-acre plot measures about 209 feet by 209 feet; a circular acre has a radius of 117.75 feet.
  2. a measure of area equal to 43,560 square feet (4,046.87 square meters). One square mile equals 640 acres.

- **Acre-Foot (AF)**
  the volume of water that will cover one acre to a depth of 1 foot.

- **actinopod**
  a type of pseudopod, needle-shaped, with internal skeleton formed from arrays of microtubules.

- **active falling area**
  the area within two tree-length radius of where a faller or a mechanised falling machine is operating.

- **active floodplain**
  the level area with alluvial soils adjacent to streams that is flooded by stream water on a periodic basis and is at the same elevation as areas showing evidence of flood channels free of terrestrial vegetation, recently rafted debris or fluvial sediments newly deposited on the surface of the forest floor or suspended on trees or vegetation, or recent scarring of trees by material moved by flood waters.

- **acute injury**
injury, usually involving necrosis, which develops within several hours to a few days after a high dose exposure to a pollutant; expressed as fleck, scorch, bifacial necrosis, etc.

- **adaptive management**
1. adaptive management rigorously combines managgment, research, monitoring, and means of changing practices so that credible information is gained and management activities are modified by experience.
2. driving forest management with scientific research. This requires excellent monitoring, and feedback loops to quickly put new knowledge into practice on the ground.

- **adaptive management areas**
landscape units designated for development and testing of technical and social approaches to achieving desired ecological, economic and other social objectives.

- **additive effects**
effects on biota of stress imposed by one mechanism, contributed from more than one source (e.g., sediment-related stress on fish imposed by sediment derived from streambank sources and from land surface sources). See also **cumulative effects**.

- **adoral**
relating to position: toward the mouth.

- **advanced regeneration**
trees that have become established naturally under a mature forest canopy and are capable of becoming the next crop after the mature crop is removed.

- **adverse (grade)**
ascending grade in the direction of travel.

- **adverse slope**
an uphill incline for hauling or skidding of logs or other loads.

- **aecial stage (aecium)**
a spore stage of the rust fungi; a cuplike structure bearing aeciospores.

- **aerial logging**
a logging system which fully suspends the logs such as done by helicopters or balloons. Not to be confused

with cable systems which use cables and supports.

- **aerial photography**
photos taken from the air at regular, spatial intervals and used in photo interpretation to provide much information about forests and landforms.

- **aerial**
pertaining to the air - as opposed to the land (terrestrial) or the water (asuatic).

- **aesthetics**
1. sensitivity to or appreciation of the forest's beauty through recognition of its unique and varied components.
2. beauty through an orderly appearance.
3. forest value, rooted in beauty and visual appreciation, affording inspiration, contributing to the arts and providing a special quality of life.

- **afforestation**
1. the establishment of trees on an area that has lacked forest cover for a very long time or has never been forested.
2. establishment of forest crops by artificial methods, such as planting or sowing on land where trees have never grown.
3. growing or planting of new woods or forests.
4. the establishment of forest trees by planting or seeding an area not previously forested.

- **age**
means age of the trees comprising a forest, crop or stand. In forests, the mean age of dominant (and sometimes codominant) trees is taken. The plantation age is generally taken from the year the plantation was begun, without adding the age of the nursery stock-Of a tree: the time elapsed since the germination of the seed, or the budding of the sprout or cutting from which the tree developed.

- **age class**
1. any interval into which the age range of trees, for-

ests, stands or forest types is divided for classification. Forest inventories commonly group trees into 20-year age classes.
2. one of the intervals, commonly 10 or 20 years, into which the age range of tree crops is divided for classification or use. Also pertains to the trees included in such an interval. For example, trees ranging in age from 21 to 40 years fall into a 30-year age class; 30 designates the midpoint of the 20-year interval from 21 to 40 years.

- **aggradation**
accumulation of sediment in a stream channel on an alluvial fan or on a floodplain. Also applied to sediment accumulation on slopes.

- **aggregated retention**
retaining trees in patches throughout a cutblock or cutting unit.

- **agricultural land**
areas used primarily for production of food and/or fibre (excludes wood fibre). Examples are croplant, pasture, orchards, vineyards, nurseries, confined feeding areas, farmsteads and ranch headquarters.

- **agrobacterium**
a naturally occurring soil bacterium that is used to transfer genes into plants.

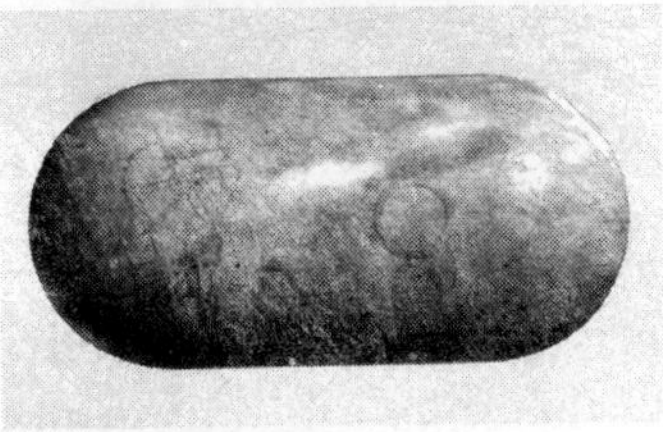

- **agroforestry**
1. the combined production of both crops and trees to maximise water efficiency.
2. the combined production of both crops and trees to maximise water efficiency.
3. cultivation system combining agriculture and forestry where trees and crops are interplanted; not common to Pennsylvania.

- **air-dried**
lumber or other wood products that have been either dried by exposure to natural atmospheric conditions outdoors or in an unheated shed or dried to equilibrium with the surrounding atmosphere. Moisture content of air-dried wood fibre depends

on relative humidity, temperature and length of drying period. Also referred to as air seasoned.

- **airtanker**

a fixed-wing aircraft fitted with tcnks and equipment for dropping suppressants or retardants.

- **alevin**

the life stage of c salmonid between egg and fry. An alevin looks like a fish with a huge pot belly which is the remaining egg sac.

- **alga**

a polyphyletic grouping of organisms with chloroplasts. Now divided into blue-green algae (eubacteria), cryptomonads, chlorarachniophytes, glaucophytes, dinoflagellates and other alveolates, euglenids (a group of Euglenozoa), various stramenopiles (also referred to as chromophytes, chrysophytes, heterokonts - including the diatoms and brown algae), haptophytes (prasinophytes), green algae (green plants) and red algae.

- **algae**

any group of lower plants having chlorophyll but no vascular system.

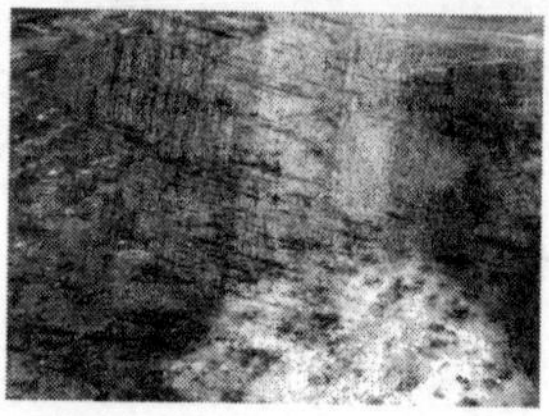

- **alkalinity**

a measure of calcium carbonate expressed as mag/l of $CaCO_3$.

- **all-aged**

forest or stand containing trees of almost all age classes up to and including trees of harvestable age.

- **all-aged or uneven-aged management**
the practice of managing a forest by periodically selecting and harvesting individual trees or groups of trees from the stand while preserving its natural appearance. This is most common in hardwood forests.

- **all-aged or uneven-aged stand**
a forest stand composed of trees of different ages and sizes.

- **allochthonous**
organic material composed of leaves and woody debris.

- **allowable cut**
1. volume of timber that may be harvested during a given period to maintain sustained production.
2. the amount of timber that may be harvested from an area over a specified period of time in accordance with the management objectives.

- **allowable cut effect**
allocation of anticipated future forest timber yields to the present allowable cut; this is employed to increase current harvest levels (especially when constrained by evenflow) by spreading anticipated future growth over all the years in the rotation.

- **aloricate**
no lorica present.

- **alternate**
leaves that are staggered or not placed directly across from each other on the twig.

- **alternate host**
one of two taxonomically different hosts required by a heteroecious rust fungus to complete its cycle; also applies to some insects.

- **alternative**
a choice between two things or among more than two things.

- **amoeba**
any of a polyphyletic grouping of organisms which

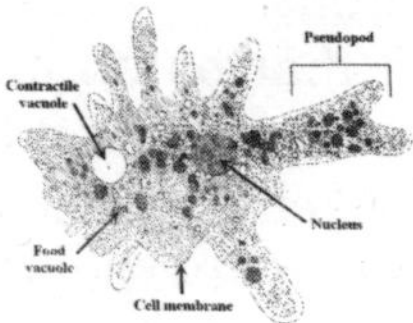

move and feed using pseudopodia.

- **amoeboid**
like an amoeba. Usually means that the organism has the capacity to produce pseudopodia.

- **amortisation**
a procedure by which the capital cost of projects, such as roads or bridges, is written off over a specified period of time as the timber volumes developed by the projects are harvested and extracted.

- **amount of water appropriated or claimed**
an entry in either cubic feet per second or acre-feet per year or gallons per day is required for all current uses.

- **anadramous**
1. fish that breed in fresh water but live their afult life in the sea. On the Pacific coast, anadramous fish include all the Pacific salmon, steelhead trout, some cutthroat trout and dolly varden char, lampreys and eulachons.
2. fish that hatch and live the first part of their lives in fresh water; then migrate to the ocean to spend their adult lives, which may be as short as six months or as long as seven years; then return to freshwater spawn.

- **analysis unit**
the basic building blocks around which inventory data and other information are organised for use in forest planning models. Typically, these involve specific tree species or type groups that are further defined by site class, geographic location or similarity of management regimes.

- **anchor**
a stump or wooden, concrete, or metal device secured in the earth to hold a line securely. Also see **deadman**.

- **anchor line**
a line to used to tie down a yarder to prevent tipping. Also see **guyline**.

- **angiosperm**
any class of flowering plants characterised by seeds that are fully enclosed by fruits.

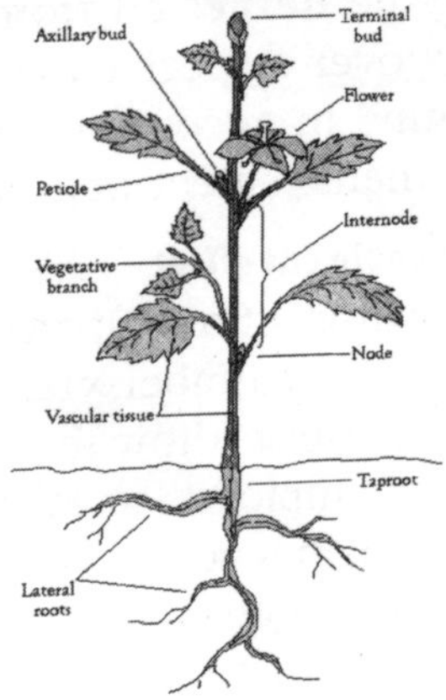

- **Animal Unit Month (AUM)**
  the amount of forage required for one month by an average animal of the genus Bos (i.e., a cow) aged 6 months or older.

- **annual**
  a plant that lives or grows for only one year or one growing season.

- **annual allowable harvest**
  quantity of timber scheduled vo be removed from a particular management unit in 1 year.

- **annual growth**
  average annual increase in the biomass of growing-stock trees of a specified area.

- **annual wildlife seed mixture**
  a mixture of soybean, millet, cow peas, sorghum, lespedeza, buckwheat and other seeds from which single-season plants are grown to serve as food or protective cover for wildlife. Some mixtures reseed naturally, while others require reseeding, light disking and fertilisation.

- **anoxic**
  an environment in which there is no free oxygen present. A situation that commonly arises in natural habitats when the biological demand for oxygen exceeds the supply, for example in sediments or when a site is organically polluted.

- **anterior**
  in front; before; opposite of posterior.

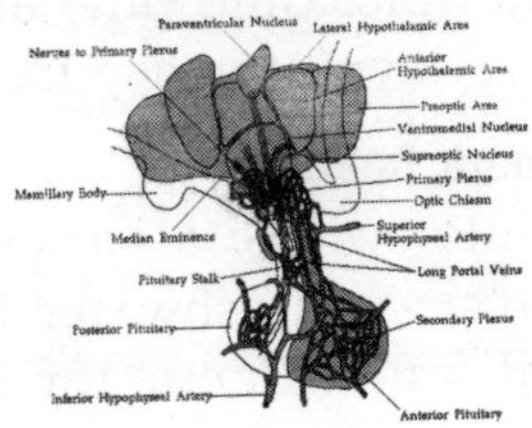

- **anthers**
  sac-like component of a flower where pollen grains are produced. The anthers open to release pollen.

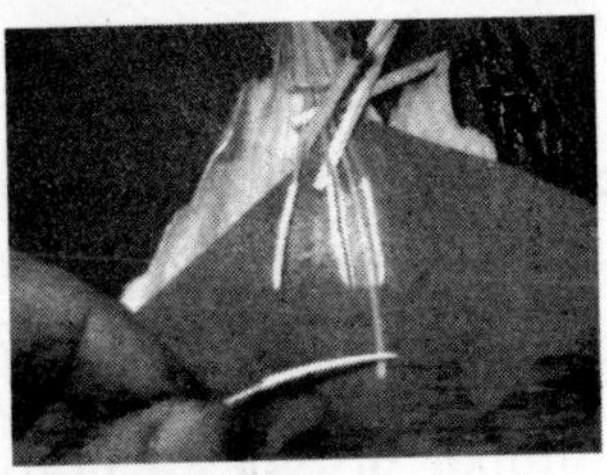

- **anthocyanins**
  1. pigments in plants responsible for pink and purple colours.
  2. presence of abnormal red-purple coloration in foliage.

- **antibiotic**
destroying or inhibiting the growth of bacteria and other microorganisms.

- **apical**
relating to the apex, the anterior pole.

- **apomorphy**
a character state derived by evolution from an ancestral state (plesiomorphy) and a novel evolutionary trait.

- **aquatic habitat**
habitat where a variety of marine or freshwater flora and fauna occur for long periods throughout the year. Examples inelude tide pools, estuaries, bogs, ponds and potential underwater diving areas.

- **aquatic habitat types**
the classification of instream habitat based on location within channel, patterns of water flow and nature of flow controlling structures. Habitat is classified into a number of types according to location within the channel, patterns of water flow, and nature of flow controlling structure. Riffles are divided into three habitat types: low gradient riffles, rapids, and cascades. Pools are divided into seven types: secondary channel pools, backward pools, trench pools, plunge pools, lateral scour pools, dammed pools and beaver ponfs. Glides, the third habitat types are intermediate in many characteristics between riffles and pools. It is recognised that as aquatic habitat types occurs in various parts of the country additional habitat types may have to be described. If that becomes necessary it will be the responsibility of the regional fishery biologist to describe and define the additional habitat types.

- **aquatic**
pertaining to water - as in aquatic habitat (ponds, lakes, streams, oceans etc).

- **aquifers**
porous, water-saturated layers of sand, gravel or bed rock through which groundwater flows that can

yield an economically significant amount of water. A large concentration of underground water.

- **arch**
  a framework through which a winchline from a skidder passes over and is used to suspend the leading ends of logs being dragged (skidded).

- **archaeological site**
  a location that contains physical evidence of past human activity and that derives its primary documentary and interpretive information through archaeological research techniques. These resources are generally associated with both the pre-contact and post-contact periods in British Columbia. These resources do not necessarily hold direct associations with living communities.

- **arched skidding**
  skidding logs with an arch to raise the front of the logs off the ground; as opposed to 'ground skidding'.

- **area regulation**
  method of controlling the annual or periodic acreage harvested from a forest, despite fluctuations in fibre-yield volumes. It leads to a managed forest.

- **area sensitive species**
  plants or animals with very specific habitat requirements that are susceptible to population decline when their habitat is altered.

- **arsenic pentavalent arsenate**
  the kind used in wood preservatives, not to be confused with commercially produced Trivalent Arsonic, which is not used in any wood preservative. Pentavalent Arsenate occurs naturally in the soil, water, air, plants and in most living creatures including humans.

- **articulated**
  with referencg to a vehicle, hinged at the centre for turning, as with a wheeled skidder.

- **artificial regeneration**
  1. establishing a new forest by planting seedlings or by direct seeding (as opposed to natural regeneration).
  2. renewal of the forest by planting or direct seeding; establishing a new stand of trees by planting seeds or

seedlings by hand or machine.

- **asexual stage**
vegetative; without sexual organs or spores.

- **aspect**
1. the direction toward which a slope faces.
2. compass direction to which a slope faces. Also called exposure.
3. the orientation of a slope with respect to the compass; the direction toward which a slope faces; north facing slopes are generally cooler than south facing slopes.

- **atmosphere**
the gaseous envelope (air) surrounding the earth.

- **attitude**
1. a manner of acting, feeling, or thinking that shows one's disposition, opinion, etc.
2. specific activity or occupancy, such as ski area, historical marker, or oil and gas lease, for which a special authorisation is issued.

- **autoecious**
completing entire life cycle on one host; especially applied to the rust fungi.

- **autotrophic**
organisms which trap energy from physical or chemical sources and use the energy to assemble the macromolecules of which

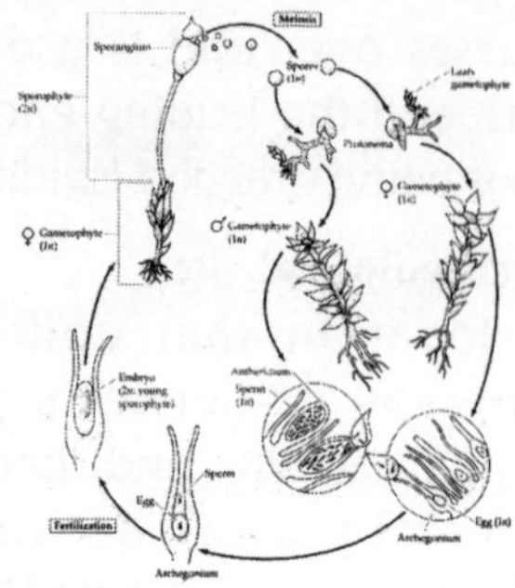

they are made. Photosynthesis is the only process by which this happens in eukaryotes, but additional processes are found among prokaryote organisms. Compare with heterotrophic.

- **available timber**
timber which is available for harvest after due recognition of constraints to protect the environment and other forest uses.

- **available volumes**
the portion of total inventory volumes that is available for harvesting after all management constraints

on timber harvesting have been considered, including definition of the timber harvesting land base, age of tree merchantability, deferrals and any other priorities or constraints on timber harvesting.

- **average long term yield**
  the annual average of the total yield over the next 200 years minus unsalvaged losses. This figure is generally greater than the long run sustained yield due to the influence of cutting old growth timber in the first few decades.

- **Average Yarding Distance**
  often abbreviated as AYD, the total yarding distance for all turns for a particular logging setting or unit divided by the total number of turns. Usually expressed as slope distance, unless otherwise stated.

- **avoidable waste**
  the volume of timber left on the harvested area that should have been removed in accordance with the utilisation standards in the cutting authority. It does not include the volume of timber that could not be removed because of physical impediments, safety considerations, or other reasons beyond the control of the licensee. Avoidable waste volumes are billed monetarily, as well as for cut control.

- **avoiders**
  plants that have little adaptability to fire.

- **axoneme**
  a geometrically packed assemblage of microtubules (subcellular scaffolding) used to support flagella, the arms of Heliozoa, etc.

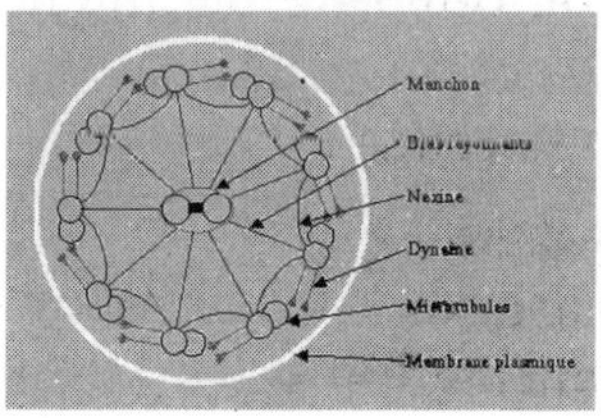

- **azimuth**
  the horizontal angle or bearing of a point measured from the true (astronomic) north. Used to refer to a compass on which the movable dial (used to read direction) is numbered in 360°.

- **back guy**
  the guy lines behind the spar, opposite the lead of the

main line or skyline in a cable logging set up. Usually applies to standing towers, rather than to swing towers.

- **backcut**

1. final cut in felling a tree. Made on the opposite side of the direction of fall.

2. the final cut in felling a tree by hand, made on the side opposite the intended direction of fall, after the undercut.

- **backfire**

blaze set in front of an advancing forest fire in an effort to check the wildfire by cutting off kts fuel supply. Controlled fire set into the wind.

- **backpack sprayer**

spray unit with plastic containers on a backpack frame. Used by individual operator to apply chemicals, such as herbicides.

- **back spar trail**

a bladed or non-bladed pathway over which mobile back spar equipment travels.

- **bacteria**

typically one-celled microorganisms which have no chlorophyll, multiply by simple division and can be seen only with a microscope

- **bacterium**

a single-cell organism that has no nucleus.

- **bacterivore**

an organism that eats bacteria.

- **banding**

a foliar symptom characterised by a limited zone of necrotic or discoloured tissue traversing the leaf, e.g., the band of tissue on a pine needle injured by SO2 or 03.

- **bank full height**
  that elevation which characterises the cross-sectional area of the active stream channel.
- **bar or blade**
  that part of the chainsaw upon which the cutting chain moves.
- **barber chair**
  1. high slab-like splint, resembling a chair back, left standing on a stump above the undercut as a result of faulty felling or heavy lean of the tree.
  2. a vertical split in a tree, generally caused by an insufficient undercut or by neglecting to cut the sapwood on both sides of a heavy leaning tree before felling. Results in a stump which looks like a high-backed chair. Very dangerous to the faller.

- **bareroot seedling**
  1. stock whose roots are exposed at the time of planting (as opposed to container or plug seedlings). Seedlings are grown in nursery seedbeds and lifted from the soil in which they are grown to be planted in the field.
  2. tree seedling grown in a nursery bed. When large enough for transplanting, the seedling is lifted from the nursery bed and the dirt is removed from the roots before packaging.
- **bark**
  1. the outside covering of the stems and roots of trees and woody plants.
  2. outward covering of the tree.

- **bark beetles**
small, cylindrical beetle of the family Scolytidae, the adult of which bores into and beneath the bark of various trees for the purpose of egg laying.

- **bark residue**
refers to the bark removed from a log and also to portions of wood and foreign matter such as sand, grit, or stones that may be imbedded in the bark.

- **bark thickness**
a measure of the vhickness of the bark at d.b.h., unless otherwise specified. Radial bark thickness is determined at a level slightly (1") below the d.b.h. to prevent callousing. It is measured from the inside of the cambium layer to the outside of the exterior bark.

- **barking drum**
large drum in which logs or billets are tumbled by mechanical rotation, the bark being removed by abrasive action.

- **barking iron**
tool with a narrow-shaped, curved blade used in removing bark by hand. Also called a spud.

- **barren land**
argas of limited plant life. Examples are mud flows, talus slopes, beaches, dunes, dry salt-flats, bare rock, surface mines and glaciers.

- **barriers**
obstructions to pedestrians, horse, or vehicular traffic. They are intended to restrict such traffic to a specific location.

- **basal area**
  1. the cross section area of the stem or stems of a plant or of all plants in a stand, generally expressed as square units per unit area. Tree basal is used to determine percent stocking. For shrubs and herbs it is used to determine phytomass. Grasses, forbs and shrubs usually measured at or less then 1 inch above soil level. Trees-the cross section area of a tree stem in square feet commonly measured at breast height (4.5' above ground) and inclusive of bark, usually computed by using d.b.h. or tallied through the use of basal area factor angle gauge.
  2. a forest measurement used to indicate stand stocking. For example, a well stocked pine stand might contain 80 to 120 square feet of basal area per acre.
  3. a measurement of the cross-sectional area of a tree trunk in square feet at breast height. Basal area (BA) of a forest stand is the sum of the basal areas of the individual trees, and is reported as BA per acre.

- **basal area factor**
  number of units of basal area per acre (or per hectare) represented by each tree.

- **basal area per hectare**
  the area of the cross-section of tree stems near their base, generally at breast height and including bark, measured over 1 ha of land.

- **base**
  where the point at which the leaf is joined to the stem.

- **base case**
  the current socio-economic conditions related to the existing forest land management strategy and the expected socio-economic conditions if the strategy remains unchanged.

- **baseflow**
  the portion of streamflow which comes from groundwater.

- **baseline information**
information collected to provide a standard against which future measurements can be compared.

- **basic silvicultural practices**
maintenance of the productivity of forest sites, restocking of denuded forest lands with commercial tree species within three years for areas west of the Coast Range and five years for areas in the Interior, protection against damage by fire, insects and diseases to predetermined standards.

- **basic silviculture**
harvesting methods and silviculture operations including seed collecting, site preparation, artificial and natural regeneration, brushing, spacing and stand tending, and other operations that are for the purpose of establishing a free growing crop of trees of a commercially valuable species and are required in a regulation, pre-harvest silviculture prescription or silviculture prescription.

- **basidial stage**
a spore stage of the rust fungi; a specialised structure in the Basidiomycetes bearing basidiospores.

- **basidiocarps**
sexual fruiting structure in the Basidiomycetes; conks, sporophore, mushrooms etc.

- **basis weight**
for linerboard and corrugating medium, the weight of the paper per 1,000 sq. ft. For fine paper, the weight, in pounds, of a ream (500 sheets) cut to its basic size, e.g., 25" x 38".

- **bearing**
a direction on the ground or on a map defined by the angle measured from some reference direction: this may be true (geographic) north, magnetic north or grid north.

- **beaver ponds**
  water impounded wither by temporary dams constructed by bank dwelling beaver or by permanent dams built by lodge dwelling beaver.

- **beaver-tailing**
  burying the whole bar of a chainsaw into the log or tree while cutting.

- **bed**
  the intended position in which a tree will be felled. In old-growth redwood timber, a bed is created by a bulldozer from piled up earth and slash so as to cushion the fall of these brittle giants.

- **bed load**
  particulates that are transported along the channel bottom in the lower layers of streamflow by rolling and bouncing.

- **bedding**
  1. land prepared before planting in the form of small mounds. The prepared land concentrates topsoil and elevates the root zone of seedlings above temporary standing water. Fertiliser is often incorporated into the bedding.

  2. a site preparation method in which special equipment is used to concentrate surface soil and forest litter into a ridge 6 to 10 inches high on which forest seedlings are to be planted.

- **bell**
  a type of choker hook that slides on the choker wire between the two knobs.

- **belly**
  a sag in any line.

- **benefit/cost analysis**
  a technique for comparing alternate courses of action by an assessment of their direct and indirect outputs (benefits) and inputs (costs). Benefits and costs are usually defined in economic and social terms.

■ **benefits**
anything contributing to an improvement in condition; advantage; help.

■ **benthic**
associated with the benthos.

■ **benthos**
the bottom sediments of rivers, lakes, ponds, etc.

■ **berm**
a low earth fill constructed in the path of flowing water to divert its direction, or constructed to act as a counter-weight beside the road fill to reduce the risk of foundation failure.

■ **Best Management Practices (BMP)**
1. a series of forest practices thought to be the best possible for a specific region and forest type.
2. a forestry practice or combination of practices, determined by a state to be the most effective means of preventing or reducing the amount of nonpoint source pollution in order to protect streams and water quality.
3. practical and economically achievable practices for preventing or reducing nonpoint source pollution.

■ **bifacial necrosis**
death of plant tissues, extending from the upper to the lower leaf surface.

■ **bight**
1. any section of a line between the ends.
2. a band or sag in a line under tension.
3. a work area made hazardous by a line or equipment under tension.

■ **biltmore stick**
graduated stick used to estimate tree diameters when held at right angles to the axis of the stem and comparing the graduations cut by lines of sight tangential to either edge of the stem.

- **bind or bound**

compression created in a cut in a tre e or log due to uneven terrain or contact pressure from other trees or logs. This will result in the chainsaw being caught in the log or tree.

- **binder**

a wire rope placed across a load on a logging truck to secure the load. AKA 'wrapper'.

- **bioballistics**

a projectile method that uses metal particles coated with genetic material that are propelled into a cell to allow for the uptake of the genetic material.

- **biodegradable**

something that breaks down to its component parts in the environment. Few items are actually biodegradable with current landfill practices, so biodegradability is a less important product claim than recyclability. Recycling captures a used item before it gets to the landfill.

- **biodiversity**

1. the variety of genes within an animal or plant species, the variety of species in a region and the variety of ecosystems or habitats within a region.
2. the richness or variety of plants and wildlife.

- **biodiversity (biological diversity)**

the diversity of plants, animals and other living organisms in all their forms and levels of organisation, including genes, species, ecosystems and the evolutionary and functional processes that link them.

- **biogeoclimatic classification**

the delineation of biotic regions or zones on the basis

of vegetation, soils, topography and climate.

- **biogeoclimatic unit**
  part of the biogeoclimatic ecosystem classification system. The recognised units are a synthesis of climate, vegetation and soil data and defined as classes of geographically related ecosystems that are distributed within a vegetationally inferred climatic space.

- **biogeoclimatic zone**
  a geographic area having similar patterns of energy flow, vegetation and soils as a result of a broadly homogenous macroclimate.

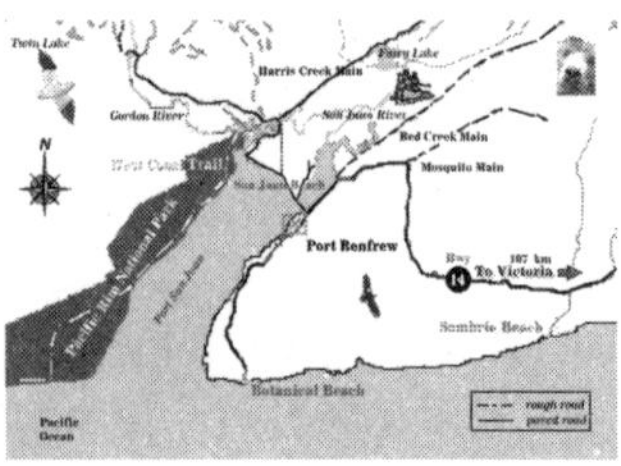

- **bioindicator species**
  species, varieties, or cultivars sufficiently sensitive to a specific pollutant to be useful as indicators for the presence of that pollutant.

- **biological control**
  the use of biotic agents such as insects, nematodes, fungi, and viruses for the control of weeds and other forest pests.

- **biological corridor**
  a habitat band linking areas of similar management and/or habitat type.

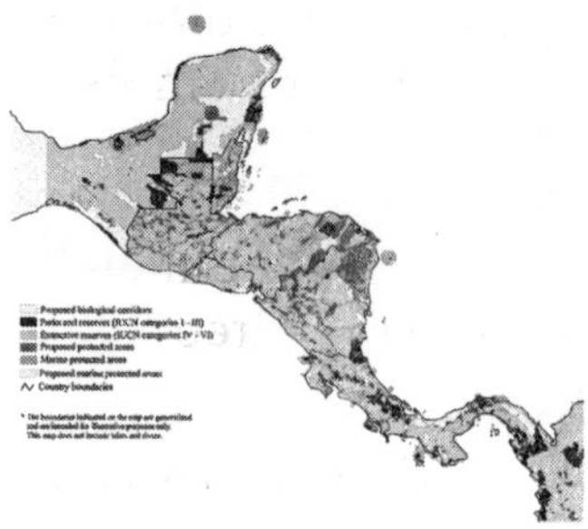

- **biological diversity**
  1. the variety of life forms in a given area. Diversity can be categorised in terms of the number of species, the variety in the area's plant and animal communities, the genetic variability of the animals, or a combination of these elements.
  2. the variety and complexity of species that are present and that interact in an ecosystem, plus the relative abundance of each.
  3. the variety of plants and animals, the communities they form, and the ecological functions they perform

at the genetic, stand, landscape and regional levels.

- **biological herbicide**
  a naturally occurring substance or organism which kills or controls undesirable vegetation. Preferred over synthetic chemicals because of reduced toxic effect on the environment.

- **biological legacies**
  features which remain on a site or landscape after a natural disturbance. These legacies include live and dead trees, coarse woody debris, soil organic matter, plants, fungi, micro-organisms and seeds.

- **biological maturity**
  the point in the life cycle of a tree at which there is no net biomass accumulation; the stage before decline when annual growth is offset by breakage and decay.

- **biological pest control methods**
  the application of whole organisms or portions of organisms as biologically sound alternatives to broad-spectrum chemical insecticides.

- **biological simplification**
  the reduction of biological diversity that results from altering the environment in ways that favour, either directly (i.e., through management) or indirectly (i.e., through pollution), certain species over many others.

- **biomass**
  1. the dry weight of all organic matter in a given ecosystem. It also refers to plant material that can be burned as fuel.

2. total woody material in a forest. Refers to both merchantable material and material left following a conventional.
3. logging operation. In the broad sense, all of the organic material on a given area; in the narrow sense, burnable vegetation to be used for fuel in a combustion system.

4. the weight of living organisms. Forests are biomass factories, producing plant, animal, and microbial biomass.

5. the total weight of all organisms in a particular population, sample, or area; biomass production may be used as an expression of site quality.

- **biomass boiler**

biomass boilers burn bark, sander dust and other wood-related scrap not usable in product production. Also called 'hogged fuel' boilers, biomass boilers make steam and heat for mill use.

- **biome**

the largest ecological unit, distinguished by an extensive complex of terrestrial communities, corresponding to a particular climatic zone or region, and associated with an environmental region such as the northern coniferous forest, the Great Plains, the tundra, or as in Pennsylvania, the eastern temperate hardwood biome.

- **bioprocessing**

integrating whole organisms or enzymes with nutrients under specific conditions to make a desired end product.

- **biosphere**

1. that part of the earth and atmosphere capable of supporting living organisms.

2. the part of the Earth and its atmosphere in which living things exist.

- **biota**

1. all living organisms of an area, taken collectively.

2. all living animals and plants.

- **biotechnology**

the collection of tools and processes used in the study of living things to make new products or services.

- **biotic pathogen**

a living organism capable of inciting disease, e.g., fungi, bacteria, viruses, etc.

- **birddog aircraft**

an aircraft carrying the person (air attack officer) who is directing fire bombing action on a wildfire.

- **birth rate**

the average number of births per year per 1,000 population at midyear. Also known as crude birth rate.

■ **black mud**
reduced muds found below the surface of sediments in lakes and rivers. The blackness is caused by the occurrence of metal sulphides.

■ **blade**
1. The part of a bulldozer or grader which digs and pushes dirt but does not lift it.

2. To scrape or level a mound of dirt or area of ground with a blade. e.g. 'It was necessary to blade the landing to smooth it for use.'

■ **bladed trail**
a constructed trail that has a width greater than 1.5 m and a mineral soil cutbank height greater than 30 cm.

■ **blaze**
to permanently mark trees, indicating those to be cut or the course of a boundary, road, or trail.

■ **bleached**
whitened; blanched.

■ **bleaching**
loss of normal colour, tending toward white, cream or tan colouration.

■ **blight**
a common term for several different diseases; usually applied to those where leaf damage is sudden and severe.

■ **blind lead**
the situation on a convex slope where the line of sight from the tail block to the yarder tower's mainline or

skyline fairlead block (bull block) is obstructed. May result in hangups to the logs being yarded if ground leading rather than full suspension is being used or available.

- **block**

  1. an area of land or timber that has been defined for management purposes. One block may be composed of stands of different species or ages.
  2. a wooden or metal case enclosing one or more pulleys used to lead a line in a specific direction, and provide with a hook, eye, or strap by which the unit may be attached to an object, such as a spar tree or anchor.

- **blowdown**

  trees felled by high winds.

- **blue-stain**

  discolouration in the sapwood of pine. At one time this was thought to be a serious defect; now it is used as high-quality interior finish.

- **blue-stain fungus**

  1. most common form of fungal stain occurring in sapwood. Conifers are most susceptible but may also occur in light-colour heartwood of perishable timbers. Commonly develops in dead trees, logs, lumber and other wood.
  2. products until the wood is dry. Reduces the grade of wood, but does not significantly reduce the strength. Some blue-stain.
  3. lumber is highly valued for specialty products.

- **Board Feet (BF)**

  lumber or timber management term. The amount of wood contained in an unfinished board 1 inch think, 12 inches long and 12 inches wide.

- **board foot**

  1. a unit of lumber measurement equal to one foot square by one inch thick.
  2. a unit of wood measuring 144 cubic inches. A 1-inch by 12-inch shelving board that is 1 foot long is equal to 1 board foot. Board foot volume is

determined by: length (feet) x width (inches) x thickness (inches)/divided by 12.

3. the standard unit of measure for standing timber, logs and lumber in the United States. A board foot is represented by a board one foot wide, one foot long and one inch thick. A unit in board feet is usually expressed in 1,000 board feet or MBF. For example, 28,500 board feet is expressed as 28.5 MBF. The number of MBF in a log or stand of timber is a function of the 'log rule' used in the local area. Different log rules will yield different estimates of MBF based on the log rule's assumptions.

4. a unit of wood 1 inch thick, 12 inches long, and 12 inches wide. One board foot contains 144 cubic inches of wood.

- **board mill**

a mill that produces boards of lumber, or sheets of chipboard, plywood etc.

- **bole**

1. trunk of a tree.
2. the main trunk of a tree.
3. the trunk of a tree, especially the merchantable or useful portion.
4. the main trunk of a tree.

- **bole length**

the length in feet of merchantable bole of trees 5.0" d.b.h. and larger between the top of a 1 foot stump and 4.0" Diameter Outside Bark (D.O.B.) unless otherwise specified.

- **bole top diameter**

the Diameter Outside Bark (D.O.B.) of the tree stem at a point on the bole above which no merchantable product section exists.

- **boreal forest**

a northern forest, as in the boreal forest Biome, characterised by evergreen conifers and long winters. The boreal forest, also referred to as a Taiga, is found in the northern parts of

North America, Europe and Asia.

■ **botanical forest products**
prescribed plants or fungi that occur naturally on Crown forest land. There are seven recognised categories: wild edible mushrooms, floral greenery, medicinal products, fruits and berries, herbs and vegetables, landscaping products and craft products.

■ **break out**
to get a log out of a hang-up or its initial lay on the ground.

■ **breaking strength**
the greatest load that a wire rope or other material can withstand before breaking. Also known as the 'ultimate strength'

■ **breast height**
the standard height, 1.3 m above ground level, at which the diameter of a standing tree is measured.

■ **broad base dip**
a surface drainage structure specifically designed to drain water from an access road while vehicles maintain normal travel speeds.

■ **broadcast burning**
a controlled burn, where the fire is intentionally ignited and allowed to proceed over a designated area within well-defined boundaries, for the reduction of fuel hazard after logging or for site preparation before planting. Also called slash burning.

■ **broadleaf**
1. a tree with leaves that are broad, flat and thin and generally shed annually.
2. having flat rather than needle like leaves; having broad leaves.

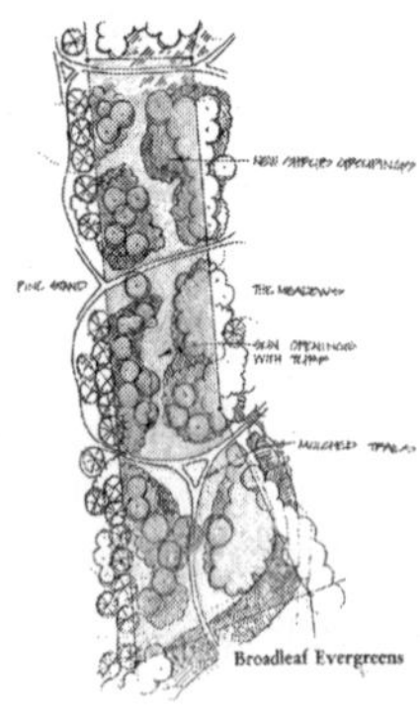

Broadleaf Evergreens

- **bronzing**

a golden brown discoloration that usually appears on the lower surface of leaves and is often an advanced stage of the silvering or glazing typical of injury by oxidants; brown colouration on needles due to spider mite infestation.

- **brow-log**

a log placed parallel to the roadway at a landing or dump. Sometimes also called a bumper-log.

- **browse**

1. shrubs, trees and herbs that provide food for wildlife.

2. portions of woody plants including twigs, shoots and leaves used as food by such animals as deer.

- **brush monkey**

usually the newest member of the cable yarding crew, usually assigned the worst or least desirable tasks, an entry-level logger.

- **brush rake**

a blade with teeth at the bottom, attached to a cat or skidder, used in mechanical site preparation. It penetrates and mixes soil and tears roots.

- **brushing**

a silviculture activity done by chemical, manual, grazing, or mechanical means to control competing forest vegetation and reduce competition for space, light, moisture and nutrients with crop trees or seedlings.

- **brush-out or swamp-out**

to clean out brush or other material around the base of trees to be felled or logs to be bucked so as to provide safe footing, prevent saw kickbacks and to provide an escape path for the faller.

- **buccal**

relating to structures associated with the mouth. From Latin bucca, mouth cavity.

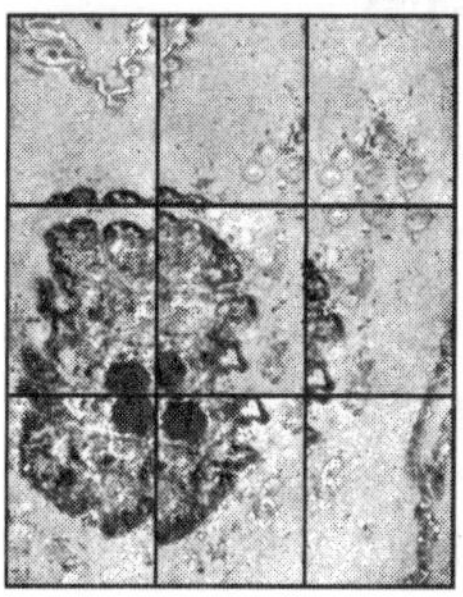

- **buck**
  to saw a felled tree into short cuts. To saw felled trees into shorter lengths.

- **bucker**
  one who saws felled trees into required lengths, such as logs, bolts or sticks.

- **bucking**
  cutting a felled tree into specified log lengths for yarding and hauling; also, making any bucking cut on logs.

- **bucksaws**
  a saw set in a frame, held on one side with both hands for cutting wood on a sawbuck.

- **buckskin**
  a log or tree with the bark fallen off.

- **bud**
  a small swelling or projection on a plant, from which a shoot, cluster of leaves or flower develops.

- **budscales**
  the protective covering of a bud.

- **budworm**
  an insect that eats buds and foliage of trees.

- **buffer area**
  an area of land and/or plants around a stream or waterbody of sufficient width to lessen entrance of pollutants (fertilisers, pesticides, eroded soil and fire retardants) into a waterbody.

- **buffer strip**
  1. a strip of land (often including undisturbed vegetation) where disturbance is not allowed or is closely monitored to preserve or enhance aesthetic and other qualities along or adjacent to roads, trails, watercourses and recreation sites.

  2. a narrow zone or strip of land, trees, or vegetation bordering an area. Common examples include visual buffers, which screen the view along roads, and streamside buffers, which are used to protect water quality. Buffers may also be used to prevent the spread of forest pests.

3. a vegetation strip or management zone of varying size, shape, and character maintained along a stream, lake, road, recreation site, or different vegetative zone to mitigate the impacts of actions on adjacent lands, to enhance aesthetic values, or as a best management practice.

4. a barrier of permanent vegetation, either forest or other vegetation, between waterways and land uses such as agriculture or urban development, designed to intercept and filter out pollution before it reaches the surface water resource.

5. a leave strip of timber along a watercourse or lake for environmental reasons, such as shade, future large woody debris recruitment, sediment filtration, and other benefits. Usually excludes equipment entry, but often permits the logging of some trees.

- **buffer**

a strip of land where disturbances are not allowed, or are closely monitored, to preserve aesthetic and other qualities adjacent to roads, trails, waterways and recreation sites.

- **built-up land**

areas predominantly covered by man-made structures. Examples are cities, strip developments, transportation facilities, institutions and industrial areas.

- **bull block**

the mainline lead block in high lead logging.

- **bullbuck**

the foreman of a cutting crew.

- **bump knots**

to cut knots off of a log. The worker on the landing who trims up logs and cuts off missed limbs and branch stubs is often called a 'knot bumper'.

- **bumper or rub tree**

a tree in a partial cut whose position next to the skid trail will protect the residual trees from damage.

- **bunch**

to assemble several logs together to form a load for subsequent skidding or yarding.

bunk
1. a crossbeam on a truck on which the logs rest.
2. to place the empty trailer of a logging truck on the tractor unit's bunk for the trip back to the landing.

bunk log
any log resting on the bunk of a truck.

business forms
fan-fold or roll forms for computer printers.

butt hook
the heavy hook on the butt rigging to which the chokers are attached in high-lead cable logging.

butt log
first log cut above the stump. Also known as butt cut.

butt log grade
the condition of the bottom log in a sawlog tree, or estimate of potential sawtimber quality for hardwood poletimber.

butt off
to cut off a piece of a log because of a defect-To square the end of a log.

butt rigging
a system of swivels and clevises which connect the haulback and mainline together and to which butt hooks are fastened. An essential part of the high lead, cable logging system.

butt rot
decay or rot characteristically confined to the base or lower bole of a tree.

cable
actually more correctly termed a wire rope.

cable logging
a yarding system employing winches and cables from a fixed position (usually a yarder and tower of some sort).

cableway
the location or pathway of a cable suspended between elevated supports so as to

constitute a track along which log carriages can be pulled.

- **calender**

to make the surface of paper smooth by pressing it between steel rollers during manufacture.

- **caliper**

1. instrument for determining tree and log diameters by measuring their rectangular projection on a straight graduated rule via two arms at right angles to (and one of them sliding along) the rule itself.

2. sight separated by a variable baseline.
3. instrument used to measure the diameter of a tree.

- **calks**

short, hobnail-like spikes in the soles and heels of boots, designed to give secure footing while walking on logs. Also known as corks.

- **callus tissue**

a protective tissue of thin-walled cells developed on wound surfaces, often beginning at the edges of a wound.

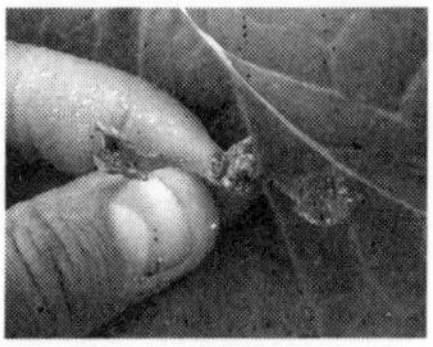

- **calyx**

the outermost whorl of sepals whose job is to protect a developing flower. It is usually green and is what we would recognise as the outside covering of a bud.

- **cambium**

1. a single layer of cells between the woody part of the tree and the bark. Division of these cells results in diameter growth of the tree

through formation of wood cells (xylem) and inner bark (phloem).

2. layer of living cells between the wood and the innermost bark of a tree. Each growing season the cambium adds anew layer of cells (by cell division) on the wood already formed, as well as a layer of inner bark on the cambium's outer face.

3. a layer of formative cells between the wood and bark in woody plants: the cells increase by division and differentiate to form new wood and bark.

4. a layer which forms across and between primary bundles where each year cells in this layer divide and grow. As the cambium divides, wood and bark cells form.

- **campfire**

a fire, not bigger than 1 m in height and 1 m in diameter, built for the purpose of cooking or providing warmth.

- **canker**

a plant disease symptom characterised by a sharply defined necrosis of cortical tissue, often sunken below bark surface.

- **canopy**

1. the forest cover of branches and foliage formed by tree crowns.

2. a layer or multiple layers of branches and foliage at the top or crown of a forest's trees.

3. the upper layer of tree crowns in a forest.

4. a layer of foliage in a forest stand. This most often refers to the uppermost layer of foliage, but it can be used to describe lower layers in a multistoried stand. Leaves, branches and vegetation that are above ground and/or water that provide shade and cover for fish and wildlife.

5. the forest cover of leaves, branches, and foliage formed by tree crowns. There may be several canopy layers.

6. the upper level of a forest, consisting of branches and leaves of taller trees. A canopy is complete (or has

100 percent cover) if the ground is completely hidden when viewed from above the trees.

- **canopy closure**

1. the progressive reduction of space between crowns as they spread laterally, increasing canopy cover.
2. the degree to which the canopy (forest layers above one's head) blocks sunlight or obscures the sky.
3. the percent of a fixed area covered by the crown of an individual plant species or delimited by the vertical projection of its outermost perimeter; small openings in the crown are included. Used to express the relative importance of individual species within a vegetation community or to express the canopy cover of woody species. Canopy cover may be used as a measure of land cover change or trend and is often used for wildlife habitat evaluations.

- **cant hook**

a pole with a movable hooked arm at or near one end, for catching hold of logs and rolling them

- **capability mapping**

a habitat interpretation for a species which describes the greatest potential of a habitat to support that species. Habitat potential may not be reflected by the present habitat condition or successional stage.

- **capital gains**

profit on the sale of an asset such as timber, land or other property. Reporting timber sales as capital gains provides certain tax advantages over reporting revenues as ordinary income.

- **carbon balance**

the concentration of carbon released into the atmosphere compared to the amounts stored in the oceans, soil and vegetation.

- **cardboard**

a thin, stiff paperboard made of pressed paper pulp or sheets of paper pasted together. Used for playing cards, greeting cards, etc. Corrugated containers are not made of cardboard.

- **carotenoids**

pigments in plants responsible for yellow and orange colours.

- **carpel**
  the female part of the centre whorl of a flower. Also known as the pistil.

- **carriage**
  a load-carrying device from which logs are suspended and which rides back and forth along the skyline on sheaves for yarding. Also called the 'skyline carriage'.

- **carrying capacity**
  1. the average number of livestock and/or wildlife that can be sustained on a management unit, compatible with management objectives for the unit. It is a function of site characteristics, management goals and management intensity.
  2. the maximum amount of animal or plant life that a particular forest environment can support indefinitely without ecosystem degradation, given the limitations of food, shelter, competition, predation and other available resources; usually expressed in terms of an individual species.

- **cascades**
  the steepest of riffle habitats. Unlike rapids, which have an even gradient, cascades consist of a series of small steps of alternating small waterfalls and shallow pools. The usual substrate of cascades is bedrock or an accumulation of boulders; however, this habitat type is occasionally found on the downstream face of woody debris dams.

- **cat**
  often short for Caterpillar tractor, or any other brand of bulldozer-type tractor. May refer to a skidding tractor or a earthmoving bulldozer.

- **cat face**
  deformed tree trunk surface usually caused by fire, disease or rot.

- **cat road**
  a path created by tractors for skidding logs, or any crude roadway or path cut by a bulldozer.

- **cat skinner**
  the person who operates a cat.

- **catchment basin**
  a hole dug adjacent to a culvert inlet to allow coarser particles to settle out.

- **catenary curve**
the shape assumed by a cable suspended freely between two points.

- **cation exchange capacity**
the sum of exchangeable cations that a soil, soil constituent, or other material can adsorb at a specific pH. Cation exchange capacity is a laboratory-analysed value

- **cell**
the most basic unit of a living organism capable of independent growth and reproduction.

- **cellulose**
1. this is the primary material of wood cell walls and it is 100% non-soluble glucose. Cellulose is a polymer chain carbohydrate (i.e. chains of sugar molecules, or polysaccharide) and is the main constituent of the cell walls of all plants, accounting for about 50% of the dry weight of wood. Also, the basic raw material (chemical compound) used in papermaking. It is the most widely distributed organic (carbon-based) compound. It is not water-soluble and is extracted by pulping.
2. the chief substance composing the cell walls or fibres of all plant tissue.

- **cellulose polysaccharide**
compound used to make walls (normally around the outside) of certain types of cell.

- **certification**
third party certification might reassure the end consumer of wood products that the wood was produced and processed in a 'sustainable' manner, without damaging forest ecosystems. Certification tells the consumer that the wood is coming from a well managed or 'sustainable' forest operation where good stewardship is practiced.

- **chain**
a measuring tape, often nylon, 50 m or 75 m in length, used to measure distances. This term is derived from an old unit of measurement: (80 Ch=1 mile).

- **chain and board**
a graphical method of calculating skyline load path and deflection by pinning a length of light chain (e.g.-lamp pull chain) to a draw-

ing board covered with grided cross-section paper on which cableway ground profiles have been drawn to scale.

- **chance**
  a logging show or operation.

- **change roads**
  to move rigging and running lines to progressively yard logs from the next area in the felled and bucked timber. AKA 'line change'.

- **channel depth**
  the average depth of a stream channel from mean high water mark to mean high water mark. Categorisation of the entrenchment and confinement is accomplished by visual analysis or by aerial photos.

- **channel gradient**
  the slope of the stream channel expressed on a percent of rise per unit length. A measure of the drop in water surface elevation per unit length of channel. Channel gradient is a parameter used in model building, channel hydraulics and flow response water yield, water use, instream and flood hazard.

- **channel roughness**
  a roughness coefficient to determine stream flow velocity and used to determine energy losses and velocities of natural stream channels.

- **channel stability rating**
  a rating of a stream channels resistance capacity to the detachment of bed and bank materials.

- **channel substrate**
  the composition of the channel substrate (stream channel bed materials).

- **characteristic visual landscape**
  the naturally appearing landscape within a scene or scenes being viewed.

- **charcoal**
  an amorphous form of carbon produced by partially burning or oxidising wood or other organic matter in large kilns from which air is excluded: used as a fuel, filter, gas absorbent, etc.

- **chase**
  to unhook chokers from the logs at the landing. The person doing this is called a 'chaser'.

▪ **chemistry, atmospheric**
the chemical composition of ambient air used to understand the degree of air pollution impact occurring on public lands and to determine the type of resource management recommendations to be made to regulators.

▪ **chemistry, pH dry deposition**
the pH of particles and aerosols deposited at the surface. Used to judge the impact of air pollutants on living and non-living public land resources.

▪ **chemistry, pH wet deposition**
the pH of precipitation used to judge the impact of acid precipitation on living and non-living public land resources.

▪ **chemistry, snowpack**
chemical composition of undisturbed accumulated snow.

▪ **chemistry, water**
this variable includes all the chemical constituents of water, nutrients, trace metals, and other organics and inorganics. Used to measure and evaluate suitability of water for various beneficial uses.

▪ **chewing insects**
insects that consume all tissues of leaves or portions of leaves, using robust mandibles for chewing.

▪ **chip**
1. small piece of wood used to make pulp. Chips are made either from wood waste in a sawmill or pulpwood operation, or from pulpwood specifically cut for this purpose. Chips are larger and coarser than sawdust.
2. to mechanically reduce logs or whole trees to small pieces for fuel, pulp, or chipboard manufacture. The equipment is called a chipper. See '**clean chips**' and '**dirty chips**'. May also be called 'in-the-woods' chipping.

▪ **chip unit**
chip volume equal to 1 cord of pulpwood.

▪ **chipboard**
a paperboard, thicker than cardboard, used for backing sheets on padded writing paper, partitions within boxes, shoeboxes, etc.

▪ **chip-n-saw**
a cutting method used in cutting lumber from trees

that measure between 6 and 14 inches diameter at breast height. The process chips off the rounded outer layer of a log before sawing the remaining cant or rectangular inside section into lumber. Chip-n-saw mills provide a market for trees larger than pulpwood and smaller than sawtimber.

- **chlorophyll**

1. the green substance in plants which in the presence of sunlight manufactures food.
2. the green pigment in plants that absorbs energy from sunlight necessary for photosynthesis.
3. a family of pigments used in photosynthesis to trap radiant energy. Normally located with chloroplasts. Chloroplasts with chlorophyll b have a bright green colour, those with chlorophylls a and c are off-green or yellow.

- **chloroplast**

1. a chlorophyll-bearing body generally found inside a leaf cell.
2. the organelle in the cytoplasm of plant cells where chlorophyll is stored.

- **chlorosis**

1. blanched or yellowish colouring in plants caused by nutrient or light deficiency.
2. yellowing of plant tissue due to failure of chlorophyll synthesis or to chlorophyll destruction.

- **chlorotic dwarf**

an abiotic disease of Pinus strobus characterised by reduced growth, chlorosis and mottling of the needles, and premature abscission of all but current needles.

- **choked**

condition in which a log is attached to a skidding unit by means of a wire rope or chain choker.

- **choker**

  1. a noose of wire rope used for skidding or yarding logs.

  2. short length of flexible wire, rope, or chain used to attach logs to a winch line or directly to a tractor.

  3. noose of wire rope for hauling a log . Short length of wire rope that forms a noose around the end of a log to be skidded and is attached to the skidding vehicle or to the butt rigging in a wire rope logging system.

- **choker setter**

  the person who puts a choker around a log and attaches it to the skidding or yarding equipment.

- **chord**

  a straight line between skyline support points.

- **chord slope**

  the slope, in percent or degrees, of the skyline chord.

- **Chromated Copper Arsenate (CCA)**

  a water-borne solution most commonly used on lumber, plywood, and timbers. This preservative protects against insects (primarily termites), fungi, and marine borers. CCA is the most widely used of all pressure-treating wood preservatives.

- **chromist**

  a term used variously to refer to some or all of those algae with chloroplasts having chlorophylls a and c (i.e. stramenopiles, cryptomonads and haptophytes). The cluster is probably polyphyletic.

- **chromosome**

  a long strand of eukaryotic DNA, often complexed with specialised proteins called histones.

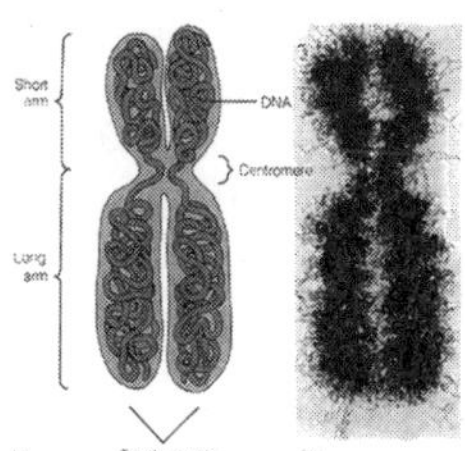

- **chronic injury**

  injury which develops after long-term or repeated low

dose exposure to an air pollutant expressed as chlorosis, bronzing, premature senescence, reduced growth, etc.

- **chute**
  the area at the landing where logs are landed on a high-lead or cable logging operation.

- **cilium (cilia)**
  a behavioural type of eukaryotic flagellum, distinctive because they occur in large numbers, have a co-ordinated behaviour, and usually direct fluids parallel to the surface.

- **CIS**
  Consumer Information Sheet, guidelines for handling treated wood, available to the consumer at retail outlets where treated lumber is sold.

- **clade**
  a single complete branch of the Tree of Life; more formally, a monophyletic group of organisms.

- **clambunk**
  a specialised forwarder designed to skid tree-length logs.

- **clamped carriage**
  a skyline carriage which clamps to the skyline when stopped and picking up logs. Clamped carriages produce more severe loading condition on the skyline than do unclamped carriages.

- **classic old growth**
  forest stands with unusually old and large trees that also meet criteria for old-growth forest.

- **clay**
  soil that is very hard when dry, but can be shaped when wet; most often found near a water source.

- **clean chips**
  wood chips for pulp manufacture without any bark

included, requires a chipper with a debarker.

- **cleaning**
  a release treatment made in a stand not past the sapling stage to free the favoured trees from less desirable species of the same age that overtop them or are likely to do so.

- **clearcut**
  an area of forest land from which all merchantable trees have recently been harvested.

- **clear-cut harvest**
  a harvesting and regeneration method that removes all trees within a given area. Clear-cutting is most commonly used in pine and hardwood forests, which require full sunlight to regenerate and grow efficiently.

- **clearcutting**
  1. a regeneration method of timber harvesting in which all suitable trees within a designated area are removed, leaving ground material in place, along with stumps and leftover woody debris. This method typically is needed to successfully re-generate most pine species, and some hardwood species of trees. Note that clearcutting should not be confused with land clearing operations, in which all material is scraped off the ground and stumps are removed to allow for a conversion of use to non-forestry purposes.
  2. in a clearcut, all the trees in a block are harvested so that a new crop of sun-loving (i.e. shade intolerant) trees can get off to a fast start. Clearcutting also facilitates mechanical planting. In some stands of trees a clearcut follows one or two 'selective cuts' which thin out the trees that compete for sunlight, water, and nutrients. Today, clearcuts are smaller in size and are designed to blend in with the landscape. Many animals, such as deer, depend on clearcuts which provide excellent forage. Clearcuts are prohibited next to streams

and lakes which require buffer zones with selective harvesting, if any.
3. a silviculture system in which an entire crop of trees is removed at one time and the area is then planted or regenerates naturally.
4. a harvesting and regeneration technique that removes all the trees, regardless of size, on an area in one operation. Clearcutting is most often used with species like aspen or black cherry, which require full sunlight to reproduce and grow well, or to create specific habitat for certain wildlife species. Clearcutting produces an even-aged forest stand.

- **clearcutting silvicultural system**

defined as a silvicultural system that removes an entire stand of trees from an area of one hectare or more, and greater that two tree heights in width, in a single harvesting operation. A new even-aged stand is obtained by planting, natural or advanced regeneration or direct seeding. The opening size and dimensions created are generally large enough to limit significant microclimatic influence from the surrounding stand.

- **clearcutting with reserves**

a variation of the clearcut silvicultural system in which trees are retained, either uniformly or in small groups, for purposes other than regeneration.

- **climate type**

the classified prevailing weather condition of a region.

- **climax**

the culminating stage in plant succession for a given site where the vegetation has reached a highly stable condition.

- **climax community**

a relatively stable and undisturbed plant community that has evolved through stages and adapted to its environment.

- **climax forest**

1. a forest community that represents the final stage of natural forest succession for its environment.
2. a plant community dominated by trees representing

the culminating stage of natural succession for that specific locality and environment.

- **climax species**
plant species that will remain essentially unchanged in terms of species composition for as long as the site remains undisturbed.

- **climbing spurs**
irons with sharp side spurs, strapped to the legs at the ankle and below the knee. Used by riggers to climb trees for topping and rigging.

- **clinometer**
1. hand instrument used by foresters and timber cruisers to measure vertical angles. Such angles, when correlated with specific distances, indicate the height of standing trees.
2. a device used in measuring the angle of elevation from eye level to the top of a tree. This instrument is used to help calculate tree height.

- **clonal lines**
a group of plants originating from buds or cuttings from the same individual.

- **clone**
a plant which is genetically identical to the parent plant. Produced asexually, e.g., from cuttings or suckers.

- **close utilisation**
maximum stump height of 30 cm; minimum top dib of 10 cm.

- **closed canopy**
the description given to a stand when the crowns of the main level of trees forming the canopy are touching and intermingled so that light cannot reach the forest floor directly.

- **closed-loop system**
when a used product is recycled into a similar product: used boxes become linerboard for making more boxes. This system helps to assure that the characteristics of the paper are maintained so that it can be recycled repeatedly.

- **closing line**
a line used to close a grapple or tongs.

- **coarse filter approach**
an approach to maintaining biodiversity that involves maintaining a diversity of

structures within stands and a diversity of ecosystems across the landscape. The intent is to meet most of the habitat requirements of most of the native species.

- **Coarse Woody Debris (CWD)**
1. sound and rotting logs and stumps that provide habitat for plants, animals, and insects and a source of nutrients for soil development.
2. portion of a tree that has falled or been cut and left in the woods. Usually refers to pieces at least 20 inches in diameter.

- **coated**
a paper covered with a suspension of china clay, pigment and adhesive.

- **coated**
a paper covered with a suspension of china clay, pigment and adhesive.

- **coccoid**
rounded in shape, ball-like.

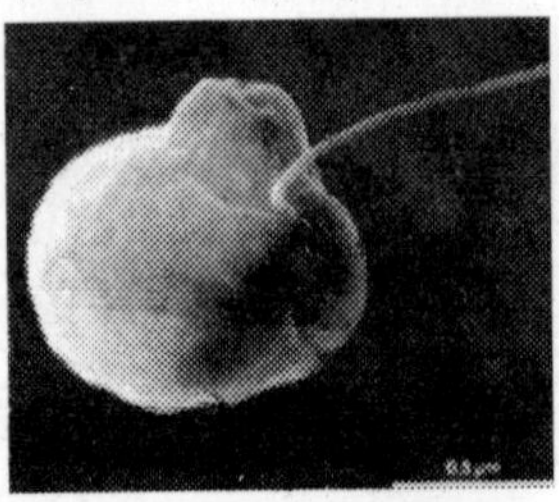

- **cocoon**
a covering, composed partly or wholly of silk or other sticky fibre, spun or constructed by many larvae as a protection for the pupal stage.

- **codominant**
in stands with a closed canopy, those trees whose crowns form the general level of the canopy and receive full light from above, but comparatively little from the sides. In young stands, those trees with above average height growth.

- **codominant trees**
trees or shrubs with crowns receiving full light from above, but comparatively little from the sides. Crowns usually form the general level of the canopy. (In stagnated stands will be small-sized and crowded on the sides).

- **cogeneration**
the process of burning fuel to produce electricity and usable steam.

- **cold deck**
a stack of logs left for later transportation (in the

woods), or a deck of logs at the mill for winter use.

- **commensal**
  an organism that derives nourishment or shelter by living in close association with another organism (the host), without damaging the host.

- **commercial lands**
  forest lands capable of bearing timber for sale.

- **commercial thinning**
  1. a silviculture treatment that 'thins' out an overstocked stand by removing trees that are large enough to be sold as products such as poles or fence posts. It is carried out to improve the health and growth rate of the remaining crop trees.
  2. partial harvesting of a stand of trees for economic gains from the harvested trees and to accelerate the growth of the trees left standing.

- **common law**
  most groundwater use is based on common law, which holds that the subsurface water belongs to whoever owns the land above it. This means landowners can withdraw as much as they want to use on their land.

- **community**
  1. all organisms in a particular habitat that are bound together by food chains and other interrelationships.
  2. a group of plants and animals living in a region or location under relatively similar conditions.
  3. a collection of living organisms in a defined area that function together in an organised system through which energy, nutrients and water cycle.

- **compartment**
  a geographic unit defined for the purposes of forest administration and inventory. The boundaries follow permanent physical features or legal demarcation where appropriate.

- **compass**
  instrument used to determine the direction of magnetic north. See **bearing** and **azimuth**.

- **competing vegetation**
  vegetation that seeks and uses the limited common resources (space, light, water and nutrients) of a forest site

needed by preferred trees for survival and growth.

- **competition**
the struggle between trees to obtain sunlight, nutrients, water, and growing space. Every part of the tree- from the roots to the crown- competes for space and food.

- **competitive**
involving or based on competition (striving for the same object, position, etc.).

- **component method**
a method of estimating or projecting population change (fertility, mortality and migration are used).

- **composition**
the proportion of each tree species in a stand expressed as a percentage of either the total number, basal area or volume of all tree species in the stand.

- **compound**
a leaf whose blade is divided into distinct leaflets.

- **concentration yard**
pulpwood yard providing facilities for unloading trucks, storage and loading for shipment.

- **condensation**
the formation of a liquid from a vapour.

- **cone**
a fruit with overlapping scales in which seeds are formed.

- **cone rake**
a device for collecting cones from a standing tree. It is lowered, usually from a helicopter, over the crown of a tree. Cones or cone-bearing branches are removed and retrieved by the machine.

- **conifer**
1. cone-bearing trees having needles or scale-like leaves, usually evergreen, and producing wood known commercially as 'softwoods'.
2. tree that is a gymnosperm, usually evergreen, with cones and needle-shaped or scalelike leaves, producing wood known commercially as softwood.
3. a tree belonging to the order Gymnospermae, com-

prising a wide range of trees that are mostly evergreens. Conifers bear cones (hence, coniferous) and needle-shaped or scalelike leaves.

- **conifer release**
  to 'release' established coniferous trees from a situation in which they have been suppressed by thinning out undesirable trees and shrubs which have overtopped them. Carried out to improve the growth of the coniferous trees released. See **brushing.**

- **coniferous**
  1. pertaining to Conifers, which bear woody cones containing naked seeds.
  2. cone-bearing trees having needles or scale-like leaves, usually evergreen, and producing wood known commercially as 'softwoods.'

- **conifers**
  trees which produce their seeds in cones and have needles.

- **conjugation**
  a type of sexual event during which two cells fuse. It may or may not lead to reproduction.

- **conk**
  a hard, fruiting body containing spores of a wood-decaying fungus.

- **consensus option**
  a management option that has a broad base of community and interest group support.

- **consequences, potential**
  a component of risk rating. Potential consequences are the detrimental events that could result from a hazard event.

- **conservancy**
  the name we use for our offices around the country which provide advice and grants to private landowners, and which administer felling licences.

- **conservation**
  1. management of the human use of the biosphere so that it may yield the greatest sustainable benefit to present generations while

maintaining its potential to meet the needs and aspirations of future generations. It includes the preservation, maintenance, sustainable utilisation, restoration and enhancement of the environment.
2. protection, improvement, and wise use of natural resources according to principles that will assure utilisation of the resource to obtain the highest economic and/or social benefits.
3. the protection, improvement, and wise use of natural resources for present and future generations.
4. the sustainable use of forest resources in a manner that does not degrade the collective resource values of a region over the long term.
5. the sustainable use of forest resources in a manner that does not degrade the collective resource values of a region over the long term.
6. the wise use and management of natural resources.

- **conservation biology**
the discipline that treats the content of biodiversity, the natural processes that produce it and the techniques used to sustain it in the face of human-caused environmental disturbance.

- **Conservation Reserve Program (CRP)**
a federal program designed to remove highly erodible, marginal farmland from production thr of ough a one-time cost-sharing payment to establish trees, grass or other cover. The landowner receives a 10-year annual rental payment to maintain the cover.

- **consumers**
organisms that feed on other organisms to sustain life.

- **consumptive activities**
forest uses which involve the removal of something from the site (hunting, fishing, timber harvesting). Non-consumptive activities include hiking, bird watching and nature study.

- **container seedling**
seedling grown in small container in a controlled environment.

- **containerboard**
the components (linerboard, corrugating medium and chipboard) used

to manufacture corrugated and solid fibreboard.

- **Continuous Forest Inventory (CFI)**
  timber sampling system that provides for periodic reme-asurement of specific stands or plots of individual trees; this shows status and periodic change over time for the forest as a whole and major subdivisions therein.
- **continuous strip landing**
  a section of roadway used as a landing in a cable logging operation where the yarder, loader, and logging truck are in a line along the road. The operation often moves down the road as logging progresses. The opposite would be a 'spot landing'.
- **contour felling**
  timber felled parallel to ground contour lines on the slope.
- **contour map**
  a topographic map which portrays relief by means of lines which connect points of equal elevation.

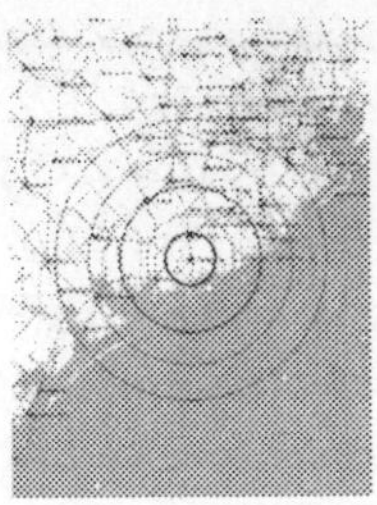

- **contraceptive prevalence rate**
  the percentage of currently married women of reproductive age (15-49) who use a method of contraception.
- **contract hauler**
  independent truck owner or a driver working for the contractor who hauls logs from the woods to the dump.
- **contract logging**
  operator doing all or part of the logging for a company-Independent logger who logs standing timber according to the terms of a contract.
- **contractual framework**
  where forest practices are primarily regulated by contracts.
- **control points**
  a system of points with established positions or elevations, or both, which are used as fixed references in positioning map features.

- **controlled burning**
  use of fire to destroy logging debris, reduce build-ups of dead and fallen timber that pose wildfire hazards, control tree diseases and clear land. Other functions of a controlled burn include clearing a buffer strip in the path of a wildfire. See **backfire**.

- **conventional ground skidding**
  any combination of rubber-tired or tracked skidding equipment.

- **conventional logging**
  any combination of mechanical or hand felling and rubber-tired or tracked skidding equipment. In the interior, cable logging is not considered conventional; on the coast, it is.

- **convergent evolution**
  evolution of similarities in unrelated groups of organisms. Adaptation for similar function may lead to novel characteristics, which are similar, although they are not inherited from a common ancestor. In some cases, such similarities may be superficial, as in the wings of birds, bats, and insects. In others, similarities can be so striking that it is difficult to determine that the traits arose independently and then later converged upon their current form.

- **Coordinated Resource Management Plan (CRMP)**
  a specific type of sub-unit plan. To date it has been used mainly for managing Crown and alienated grazing lands. This plan involves consultation with resource agencies and resource users in establishing objectives in the management and development of a specific area.

- **coppice**
  the tendency of certain tree and brush species (such as red alder and bigleaf maple) to produce a large number of shoots when a single or few stems are mechanically removed but the root system left intact.

- **cord**
  1. 128 cubic feet of stacked roundwood (whole or split, with or without bark) containing wood and airspace, with all the pieces of similar length and lined up on ap-

proximately the same direction. Example: a pile of firewood 4'x4'x8'.
2. a stack of round or split wood consisting of 128 cubic feet of wood, bark, and air space. A standard cord measures 4 feet by 4 feet by 8 feet. A face cord or short cord is 4 feet by 8 feet by any length of wood under 4 feet.
3. a unit of wood volume equal to 128 cubic foot of stacked wood. Commonly used for smaller timber utilised for firewood or pulping.

- **corduroy**
logs placed transversely along a road, usually with branches intact and covered with fill material, to float the road over soft subsoils.

- **core competencies**
the portfolio of skills that make you good at what you do. Your core competencies should remain constant and define your organisation. Core competencies include things like: teaching, marketing, trading, manufacturing and leadership.

- **critical issues**
internal or external issues which may have a severe impact on your ability to execute a strategic plan. As an example, proper incentives might encourage conservation, long range planning and ecosystem management.

- **critical wildlife habitat**
part or all of a specific place occupied by a wildlife species or a population of such species and recognised as being essential for the maintenance of the population.

- **critical winter range**
forested habitat, usually stands of mature or old-growth conifers, which provides deer and elk with resources critical to survival during severe winters.

- **crop**
the harvest of timber from a wood or forest.

- **crop tree**
1. a tree in a young stand or plantation selected to be

carried through to maturity until an interim or final harvest. any tree selected to grow to final harvest or to a selected size.
2. crop trees are selected for quality, species, size, timber potential, or wildlife value.

3. any tree selected to grow to final harvest or to a selected size. Crop trees are chosen on the basis of quality, species, size, timber potential, or wildlife value.
4. a term traditionally reserved to describe a tree of a commercially desirable species, with the potential to grow straight, tall and vigorously. However, a crop tree can be one selected for nontimber purposes (varying with landowner objectives), such as mast production or den tree potential.

- **cross pollination**

the fertilisation of an egg in one plant by a sperm cell found in a pollen grain of another plant. The resulting seed will have the hereditary characteristics from both parents.

- **cross-ditch**

a ditch excavated across the road at an angle and at a sufficient depth, with armouring as appropriate, to divert both road surface water and ditch water off or across the road.

- **cross-drain culvert**

a culvert used to carry ditch water from one side of the road to the other.

- **crossing the lead**

felling a tree at an angle across the established felling pattern.

- **crown**

1. the live branches and foliage of a tree.
2. the upper part of a tree or other woody plant that carries the main system of branches and the foliage.

- **crown class**

the relative position of the tree or shrub crown with respect to competing vegetation surrounding the tree or shrub. Crown class for each tree or shrub is judged in the context of its immediate environment; that is, those trees or shrubs which are competing for sunlight with the subject tree. Crown class is essentially a classification of competition for light and is aimed at separating trees that are growing freely from those that are not. It designates trees or shrubs with crowns of similar development and occupying similar positions in the crown canopy. This is an ocular classification of trees or shrubs based on dominance in relation to adjacent trees or brush as indicated by crown development and amount of sunlight received from above and on the sides. In uneven-aged stands of tolerant species (in which the trees are not in small even-aged groups), trees in the intermediate crown position in the stand and with medium-sized crowns will be considered comparable to codominants of even-aged stands and coded as such. An example of this would be white spruce in an aspen or birch stand. As a general rule, in multi-story stands crown class for each tree must be judged in the context of its immediate environment, that is, those trees affecting it or being affected by it in terms of crown competition. In cases where the overstory consists of scattered veterans standing above larger numbers of younger trees, a considerable portion of the understory trees will undoubtedly be classified as dominant or codominant.

- **crown closure**
  the condition when the crowns of trees touch and effectively block sunlight from reaching the forest floor.

- **crown cover**
  the degree to which the crowns of trees are nearing general contact with one another.

- **crown density**
  the amount, compactness or depth of foliage of a tree crown.

- **crown fire**
  a fire burning into the crowns of the vegetation, generally associated with an intense understory fire.

- **crown foliage density**
a visual index of the amount of foliage per unit of crown. Used for calculation of foliage structure and a measure of the severity of defoliation and disease.

- **crown form (shape)**
the configuration crown of a standing tree or shrub-circle, triangle, neolloid, parabola, rectangle or horizontal ellipse. Used to model vegetation structure and to determine foliage volume and percent growth cover by height.

- **crown length (depth)**
the vertical distance from the top of the leader to the base of the crown, measured to the lowest live branch-whorl with live branches in at least 3 quadrants, and continuous with the main crown. Used to develop horizontal-vertical profiles and biomass estimates.

- **crown ratio**
the percent of the compacted portion of the tree bole or shrub supporting green, live, healthy foliage when compared to the total length or height.

- **crown thinning**
removing superfluous live growth in a tree crown to admit light, reduce weight and lessen wind resistance.

- **crown volume percent**
the percentage of a given space occupied by live foliage.

- **crown width (diameter)**
the span of the crown of a tree or shrub.

- **cruise**
1. a survey of forestland to locate timber and estimate its volume by species, products, size, quality or other characteristics.
2. the estimate obtained from such a survey.

- **cruiser**
one who conducts surveys of timber land. Also known as an estimator.

- **crummy**
a vehicle used for crew transportation to and from the woods.

- **crust**
outer most layer of the earth comprised of soil rock.

- **cryopreservation**
the storage of living materials at very low temperatures within liquid nitrogen to protect against damage.

- **cryptic**
hidden or concealed.

- **cryptogam**
nineteenth century concept broadly covering small algae, fungi and bacteria.

- **cubic scale**
estimate of the cubic-foot volume of wood fibre in a tree, log or other wood product.

- **cull**
1. trees or logs or portions thereof that are of merchantable size but are rendered unmerchantable by defects.
2. tree or log that is unmerchantable because of defects.- Logs that are rejected or parts of logs deducted in measurement because of defect.
3. a tree or log of marketable size made useless for all but firewood or pulpwood because of shape, disease, defect, insect infestation or injury. tree or snag that does not meet merchantable specifications.
4. any log or tree which is unmerchantable due to size, defect, or damage. May also be used to describe something which is generally worthless, e.g. 'As a choker setter, he was a cull.'
5. a tree of such poor quality that it has no merchantable value in terms of the product being cut. However, a timber cull tree may have value for wildlife or aesthetics.

- **cultivated cropland**
areas that are tilled and dominated by vegetation grown for the production of food and/or fibre, provided these areas do not qualify as Treeland, Grassland, or Shrubland. Cultivated cropland includes fallow land, land in any stage of annual crop production, and land being regularly cultivated for production of crops from

perennial plants. Examples are orchards; nurseries of all types; and areas dominated by such species as wheat, barley, corn, soybeans, grapes and berries.

- **cultural diversity**

the variety and variability of human social structures, belief systems and strategies for adapting to biological situations and changes in different parts of the world.

- **cultural heritage resource**

an object, a site or the location of a traditional societal practice that is of historical, cultural or archaeological significance to the province, a community or an aboriginal people. Cultural heritage resources include archaeological sites, structural features, heritage landscape features and traditional use sites.

- **culture**

the sum of ways of living built up by a group of human beings, which is transmitted from one generation to another.

- **culvert**

1. a buried pipe that allows streams, rivers or runoff to pass under a road.
2. a conduit through which surface water can flow under roads.

- **cumulative effect**

1. the sum total of all impacts affecting anadromous fish.
2. effects on biota of stress imposed by more than one mechanism (e.g., stress in fish imposed by both elevated suspended sediments concentrations in the water and by high water temperature).

- **cunit**

1. unit of volume consisting of 100 cubic feet. Unit of measure or stacked pulpwood that equals 100 cubic feet of solid wood (does not include bark or air volume).
2. a unit of wood volume measuring 100 cubic feet. More commonly used for estimating smaller timber to be utilised for pulping or firewood. However, in the 1990s, the U.S. Forest Service began to sell all timber based on the cubic foot volume rather than MBF.

- **cut**

the excavation required to lower the natural ground

line to the desired road profile.

- **cut bank**
  the excavated bank from a ditch line to the top of the undisturbed slope of a road.

- **cut period**
  the interval between major harvesting operations in the same stand.

- **cut sheets**
  fine paper cut to letter or legal size for use in printers, copiers and fax machines.

- **cut slope**
  the face of an excavated bank required to lower the natural ground line to the desired road profile.

- **Cut Slope Ratio (CSR)**
  the slope of a cut bank on the road or landing. Usually expressed as a ratio of X:1, where X is the number of feet run per one foot of rise. Hence a CSR of 2:1 is flatter than one of 1:1. Often of importance when a swing yarder is set up on a road and the road is used as the landing.

- **cut-and-fill**
  system of bench construction on hillslopes to produce road rights-of-way and landings whereby convex slopes are excavated and concave slopes (gullies) are filled; also, excavation of the upslope side of the right-of-way, and fill on the down slope side. So called half-bench construction.

- **cutblock adjacency requirements**
  integrated resource management requirements that specify the desired spatial relationships among cutblocks.

- **cutting contract**
  a written, legally binding document used in the sale of standing timber. The contract specifies the provisions covering the expectations and desires of both buyer and seller.

- **cutting cycle**
  the planned time interval between major harvesting operations within the same stand- usually within uneven-aged stands. For example, on a 10-year cutting cycle in a hardwood stand, trees are harvested every 10 years.

■ **cutting pattern**
the visible pattern to which trees are felled during a cutting operation.

■ **cutting unit**
an area of timber designated for harvest. Several cutting units may make a timber harvest plan or a timber sale.

■ **Cut-To-Length harvester**
a self-propelled tracked or wheeled harvesting machine, designed to fell, limb, and buck a tree into logs, usually 20 feet or less in length. Most commonly the harvesting head is mounted to an articulating arm. Not to be confused with a full-tree length harvester or feller-buncher. Most commonly CTL harvesters work with forwarders which transport the bunched log piles to the landing.

■ **Cut-To-Length system (CTL)**
a feller-processor that combines felling with delimbing and bucking at the tree stump.

■ **cycle**
a complete set of operations or tasks that is repeated.

■ **cyclone cleaner**
1. a container where the pulp is spun around to remove any dirt particles.
2. part of the food ingestion structures (mouth) of some cells; usually a channel of microtubules that draws newly formed food vacuoles away from the cytostome and into the cell.

■ **cytoplasm**
the matter which makes up cells, within which organelles occur.

■ **cytoproct**
found in some ciliates, the site at which old food vacuoles fuse with the cell surface and undigested residues are excreted.

■ **cytoskeleton**
intracellular components used to provide shape to a cell or to create tracts along which cellular organelles may be moved, mostly comprised of microtubules and actin filaments.

■ **cytostome**
literally, 'the cell mouth', only used in reference to organisms which ingest food at one or more par-

ticular locations and then best used in reference to the region(s) of the cell surface through which food gains entry into the cell; part of the 'mouth' structures; see also **cytopharynx.**

- **DIB**
  Diameter Inside Bark.

- **damaged timber**
  timber that has been affected by injurious agents such as wind (as in the case of blowdown), fire, insects, or disease.

- **danger tree**
  a live or dead tree whose trunk, root system or branches have deteriorated or been damaged to such an extent as to be a potential danger to human safety.

- **dangling head**
  a type of harvesting head, usually used on a CTL harvester, with a wrist-like action and attachment to the harvester boom. Also called a 'single-grip harvesting head' and can rotate in all three axes.

- **dark reactions**
  the photosynthetic process in which food (sugar) molecules are formed from carbon dioxide from the atmosphere with the use of ATP; can occur in the dark as long as ATP is present.

- **daughter cells**
  the products of cell division of protists.

- **daughter isotope**
  an isotope that forms from the radioactive decay of a parent isotope. A daughter isotope may or may not be of the same element as its parent. If the daughter isotope is radioactive, it will eventually become the parent isotope of a new daughter isotope. The last daughter isotope to form from this process will be stable and nonradioactive.

- **day-lighting**
  a practice in which trees shading an access road are removed to increase the sunlight on the roadway and along its periphery. This relatively inexpensive practice maximises forest edge and cover for wildlife and helps maintains passable roads year-round by speeding up the drying process.

■ **DBH (Diameter At Breast Height)**

1. the diameter of a tree as measured at breast height. Standard DBH is measured at 4.5 feet above the ground.
2. the stem diameter of a tree measured at breast height, 1.3 metres above the ground.

■ **DDT (Dichloro Diphenyl Trichloroethane)**

a powerful insecticide.

■ **deactivation**

measures taken to stabilise roads and logging trails during periods of inactivity, including the control of drainage, the removal of sidecast where necessary and the re-establishment of vegetation for permanent deactivation.

■ **deadman**

a completely buried anchor, often a log, used where no suitable stumps exist for anchoring.

■ **death rate**

1. the average number of deaths per year per 1,000 population at midyear. Also known as crude death rate.
2. the ratio between deaths and individuals in a specified population at a particular time.

■ **debarker**

machine for removing bark from logs or bolts. Machine used to remove the bark from roundwood prior to processing into chips, lumber or other wood products.

■ **debris flow**

1. the rapid, downward mass movement of particles coarser than sand, often including boulders one meter or more in diameter, at a rate ranging from 2 to 40 kilometres per hour. Debris flows occur along fairly steep slopes.
2. the material that descends in such a flow. deflation The process by which wind erodes bedrock by picking up and transporting loose rock particles.

■ **deciduous plant**

1. plants characterised by a specific growth and dor-

mancy cycle, with certain parts falling at the end of the growing period, as leaves, fruits, etc., or after anthesis, as the petals of many flowers.

2. plants having leaves of this type. As contrasted with Evergreen which remains verdant throughout the year.

- **deck**

1. pile of logs on a landing. Area or platform on which wood is placed.

2. the angle between true (geographic) north and magnetic north (direction of the compass needle). Declination varies from place to place and can be 'set' on a compass for a particular location.

- **decomposer**

1. any of various organisms (as many bacteria and fungi) that feed on and break down organic substances (such as dead plants and animals).

2. organisms that digest dead organisms to sustain life.

- **decomposition**

the breakdown of matter by bacteria and fungi, changing the chemical makeup and physical appearance of materials.

- **defect**

crook, conk, decay, split, sweep or other injury that decreases the amount of usable wood that can be obtained from a log.

- **deficit forest**

a forest in which existing stands cannot provide enough harvest volume to maintain the harvest at the level of long run sustained yield until the stands created when existing stands are cut become available for harvest. See also **surplus forest**.

- **deflection**

the vertical distance between the chord and skyline, usually measured at midspan and expressed as a percentage of the span. Also known as 'sag'.

- **defoliation**

1. the destruction of plant leaves by pests.

2. an agent that damages trees by destroying leaves or needles. Deforestation: clearing an area of forest on a non-temporary basis for

another use. Clearcutting (even with stump removal), if shortly followed by reforestation for forestry purposes, is not deforesting.
3. any chewing insect that consumes the leaves or needles of plants.
4. insects that destroy foliage.- Chemicals that cause plants to drop their leaves.

- **deforestation**
1. the unintentional or intentional conversion of land use from forest to nonforest. Associated with nonrenewable timber harvesting practices in ecologically sensitive areas, such as tropical rainforests.
2. clearing an area of forest for another long-term use.

- **degradation**
1. the diminution of biological productivity or diversity.
2. the process by which a stream's gradient becomes less steep, due to the erosion of sediment from the stream bed. Such erosion generally follows a sharp reduction in the amount of sediment entering the stream.

- **degrade**
any defect that lowers the grade or quality of a log.

- **de-inking**
a series of washing and floatation processes in which water and/or soap-like chemicals called surfactants remove the ink from recycled paper.

- **deleterious substance**
any substance that, if added to water, would degrade or alter the quality of the water so that it becomes deleterious to fish or fish habitat, or becomes unsuitable for human consumption or any other purpose for which it is legally licensed (such as irrigation and livestock watering).

- **deletion**
the loss of a chromosome segment without altering the number of chromosomes.

- **delimber**
1. an excavator based machine with an attachment which removes a tree's limbs and bucks the top off the tree at the landing. A log loader then places the processed logs on a truck.

2. landing-based piece of equipment used to limb trees brought in unlimbed, usually cut with full tree length harvesters. May be of the stroke delimber type or pull-through delimber model.

- **deliquescent**
  to ramify into fine divisions, such as abnormal numbers of buds, twigs, branches or leaves, e.g., witches' broom development.

- **delpetion allowance**
  deduction from taxable income derived from wasting asset. The Internal Revenue Code of the United States permits the calculation of depletion allowances either on the basis of a percentage of the gross income from the property in question or on a per-unit-of-product basis. Depletion differs from depreciation in that the asset subject to depletion cannot be replaced; a mine or an oil field cannot be replaced in the same manner that a factory or machine can be replaced.

- **delta**
  an alluvial fan having its apex at the mouth of a stream.

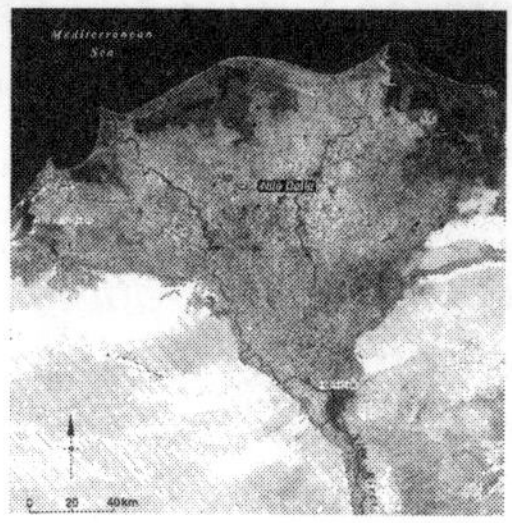

- **den tree**
  a tree with cavities in which birds, mammals, or insects such as bees may nest (also known as cavity tree).

- **dendrites**
  short, highly branched fibres that carry signals toward the cell body of a neuron.

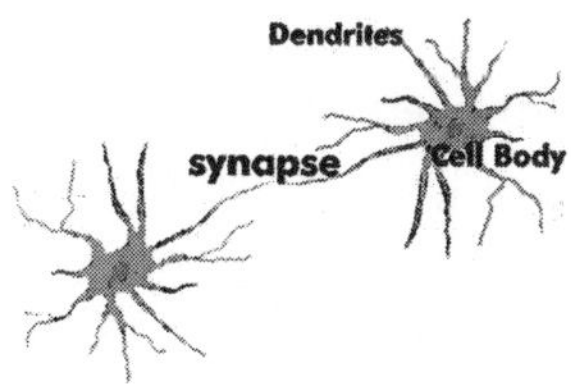

- **dendrochronology**
  the process of determining the age of a tree or wood used in structures by counting the number of annual growth rings.

- **dendrology**
  study and identification of trees.

- **density**
  number of organisms per unit of space.

- **Deoxyribo Nucleic Acid (DNA)**
a nucleic acid composed of two polynucleotide strands wound around a central axis to form a double helix; the repository of genetic information. Nucleic acid

Deoxyribonucleic Acid Molecule

- DNA is composed of 2 phosphate sugar chains that are connected by a pair of base molecules.
- Phosphates are the yellow phosphorus atoms surrounded by the red oxygen atoms.

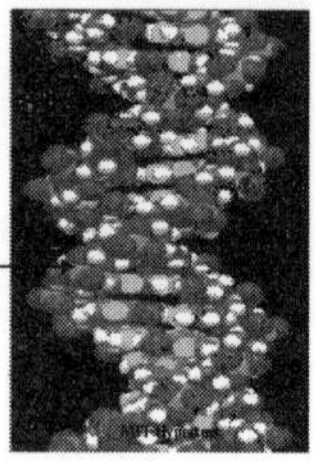

that functions as the physical carrier of inheritance for 99% of all species. The molecule is double-stranded and composed of two strands in an antiparallel and complementary arrangement. The basic unit, the nucleotide, consists of a molecule of deoxyribose sugar, a phosphate group and one of four nitrogenous bases.

- **deoxyribose**
five-carbon sugar found in nucleotides of DNA.

- **depletion**
an income tax allowance reflecting the purchase price paid for merchantable timber, usually on fee simple land. Also, a term used to refer to the process of harvesting your growing stock.

- **depreciation**
the annual reduction in value of a piece of equipment or capital asset due to wear and tear or obsolescence.

- **depth diversity gradient**
the increase in species richness with increasing water depth until about 2000 meters below the surface, where species richness begins to decline.

- **depth to bedrock or restriction**
the vertical distance from the mineral soil surface to unbroken solid rock or restriction.

- **depth to mottling or water**
mottling is the occurrence of small spots of colour which contrast with the general matrix colour of the soil. These spots of colour, or mottles, commonly appear as small spherical splotches. Mottle colours are either:
1. grey on a matrix subsoil colour of yellowish-brown, or
2. reddish-brown on a matrix subsoil colour of grey or

greyish-brown. The shorter the depth to mottling, the more poorly drained the soil.

- **dermal system**
  plant organ system that provides the covering for the plant.

- **dermis**
  one of the two layers of skin; a connective tissue layer under the epidermis containing elastic and collagen fibres, capillary networks and nerve endings.

- **desert**
  a region with an average annual rainfall of 10 inches or less and sparse vegetation, typically having thin, dry and crumbly soil. A desert has an aridity index greater than 4.0.

- **desert biome**
  characterised by dry conditions and plants and animals that have adapted to those conditions; found in areas where local or global influences block rainfall.

- **desert pavement**
  a closely packed layer of rock fragments concentrated in a layer along the Earth's surface by the deflation of finer particles.

- **desert varnish**
  a thin, shiny red-brown or black layer, principally composed of iron manganese oxides, that coats the surfaces of many exposed desert rocks.

- **desertification**
  the process through which a desert takes over a formerly non-desert area. When a region begins to undergo desertification, the new conditions typically include a significantly lowered water table, a reduced supply of surface water, increased salinity in natural waters and soils, progressive destruction of native vegetation and an accelerated rate of erosion.

- **design flaw**
  permitting standards to vary among students while keeping time constant instead of providing each student the time to reach high standards.

- **designated area**
  an identifiable geographic unit of the forest land base that requires a specific combination of forest practices to adequately protect important resource values.

- **designated skid road/skid trail**
  a pre-planned network of skid roads or skid trails, designed to reduce soil disturbance and planned for use in subsequent forestry operations in the same area. Multiple passes by tracked or rubber-tired skidders or other equipment are anticipated.

- **desired future stand condition**
  a description of the characteristics of the future stand.

- **desired plant community**
  a plant community that produces the kind, proportion and amount of vegetation necessary for meeting or exceeding the land-use plan requirements or ecological site objectives. The desired plant community must be consistent with the site's capability to produce the desired vegetation through management, land treatment or a combination of the two.

- **desmosome**
  a circular region of membrane cemented to an adjacent membrane by a molecular glue made of polysaccharides; found in tissues that undergo stretching.

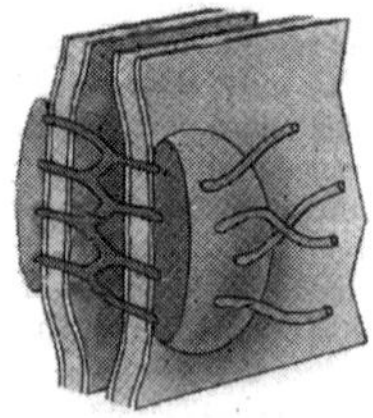

- **detrimental soil disturbance**
  changes caused by forest practices in the physical, chemical, or biological properties of the soil, including the organic forest floor and the mineral soil extending from the surface to the depth at which the unweathered parent material is encountered. Such changes may result in a loss of productive growing site, reduced site productiv-

ity, or adverse impacts on resource values.

- **detrimental soil disturbance**
  the condition where established threshold values for soil properties are exceeded and result in significant change.

- **detrital sediment**
  sediment that is composed of transported solid fragments of pre-existing igneous, sedimentary, or metamorphic rocks.

- **detritus**
  fragments of dead plant and animal material before, during and after breakdown by agents of decay. May incorporate inorganic matter (such as mud).

- **deuterostomes**
  animals in which the first opening that appears in the embryo becomes the anus while the mouth appears at the other end of the digestive system. Main groups include chordates and echinoderms.

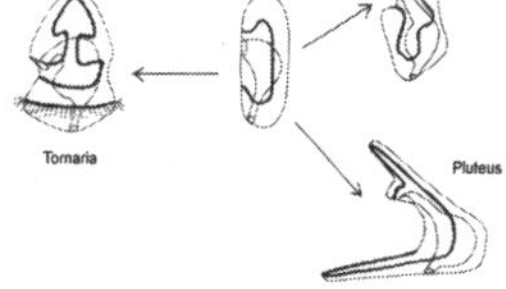

- **development**
  the advancement of the management and use of natural resources to satisfy human needs and improve the quality of human life. For development to be sustainable it must take account of social and ecological factors, as well as economic ones, of the living and non-living resource base, and of the long-term and short-term advantages and disadvantages of alternative actions.

- **development objectives**
  the short-term (often 5-year) planning objectives for a specific management area.

- **development plan**
  a specific plan outlining harvesting, road construction, protection, and silviculture activities over the short-term (often 5 years) in accordance with the approved forest management plan.

- **dewatering**
  condition in stream channel when all the water flow occurs within the permeable streambed sediments, so no surface water is left; common in small streams with

considerable accumulations of gravel.

- **diabetes mellitus, types I and II**
  a disorder associated with defects in insulin action. Type I diabetes is characterised by inadequate insulin secretion; Type II diabetes is characterised by impaired insulin secretion in response to elevated blood glucose levels or by loss of sensitivity to insulin by target cells.

- **diagnosis**
  the set of processes that cause physical and chemical changes in sediment after it has been deposited and buried under another layer of sediment. Diagnosis may culminate in lithification.

- **Diameter at Breast Height (DBH)**
  tree DBH is outside bark diameter at breast height. Breast height is defined as 4.5 feet (1.37m) above the forest floor on the uphill side of the tree. For the purposes of determining breast height, the forest floor includes the duff layer that may be present, but does not include unincorporated woody debris that may rise above the ground line.

- **diameter classes**
  classification of trees based on diameter outside bark measured at d.b.h. In forest surveys, each diameter class encompasses approximately 2 inches: the 6-inch class would include trees 5.0 through 6.9 inches in d.b.h.

- **diameter inside bark**
  diameter measurement of a standing tree or log in which the estimated or actual thickness of the barks discounted.

- **diameter limit**
  the removal of trees from a stand, based on the criterion of diameter. Generally, trees of less than a predetermined diameter are left unharvested.

- **diameter outside bark**
  measurement of tree diameter in which the bark is included.

- **diameter tape**
  a graduated tape based on the relationship of circumference to diameter which provides direct measure of tree diameter when stretched

around the outside of the tree, usually at breast height. See **DBH.**

- **diameter, basal (diameter at root collar)**
  the straight line passing through the centre of a cross section of a bole measured at the root collar of a shrub or tree.

- **diameter, stump**
  the diameter of a tree inside or outside bark at stump height (generally one foot above the ground on the uphill side).

- **diameter-limit cut**
  1. a timber harvesting treatment in which all trees over a specified diameter may be cut. Diameter-limit cuts often result in high-grading.
  2. a selection method in which all marketable trees above a specified diameter are harvested. Diameter-limited cutting can lead to long term degradation of the stand.

- **diaphragm**
  a dome-shaped muscle that separates the thoracic and abdominal cavities.

- **diastole**
  the filling of the ventricle of the heart with blood.

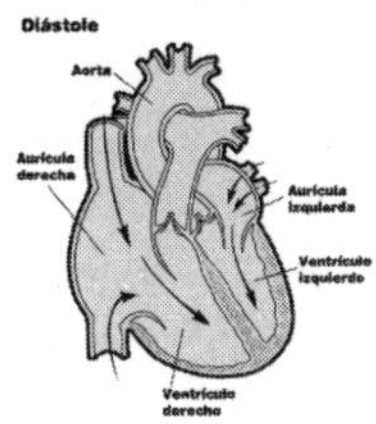

- **diatomaceous earth**
  fossilised deposits of diatoms; used for abrasives, polishes and as a filtering agent.

- **DIB (Diameter Inside Bark)**
  the diameter of a tree or log excluding bark thickness.

- **dibble**
  a tool used to make holes in the ground for planting tree seedlings.

- **dichotomous tree**
  a tree where all branching points are dichotomies. That is, a tree is dichotomous if at each branch point there are only two immediate descendents. This is in contrast to a polytomous tree.

- **dichotomy**
  a branch point on a tree that has two immediate descendents.

■ **dicots**

one of the two main types of flowering plants; characterised by having two cotyledons, floral organs arranged in cycles of four or five and leaves with reticulate veins; include trees (except conifers) and most ornamental and crop plants.

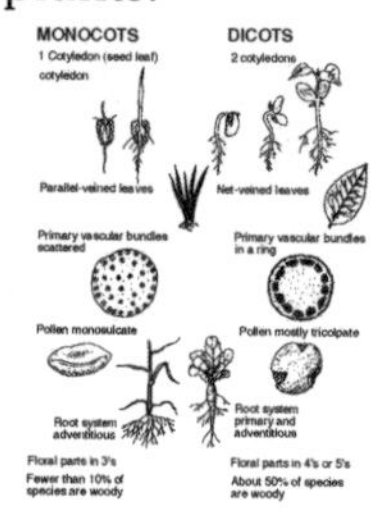

■ **dictyosomes**

organelles in plant cells composed of a series of flattened membrane sacs that sort, chemically modify, and package proteins produced on the rough endoplasmic reticulum. Also known as the Golgi Apparatus.

■ **diencephalon**

part of the forebrain; consists of the thalamus and hypothalamus.

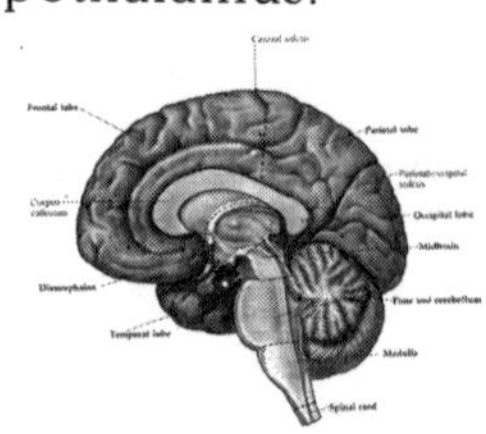

■ **difficult site**

forest sites with environmental conditions that are unfavourable for tree establishment and growth.

■ **diffusion**

the spontaneous movement of particles from an area of higher concentration to an area of lower concentration.

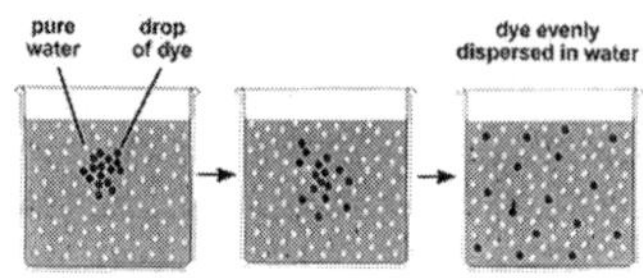

■ **diffusion feeding**

feeding strategy in which the predator relies on the movements of the prey to make contact - as in heliozoa and suctoria.

■ **digester**

a big tank that cooks wood chips and chemicals at high temperatures to separate the cellulose fibres.

■ **digestion**

the process of breaking down food into its molecular and chemical components so that these nutrient molecules can cross plasma membranes.

- **digestive system**

  one of eleven major body organ systems in animals; converts food from the external environment into nutrient molecules that can be used and stored by the body and eliminates solid wastes; involves five functions: movement, secretion, digestion, absorption and elimination.

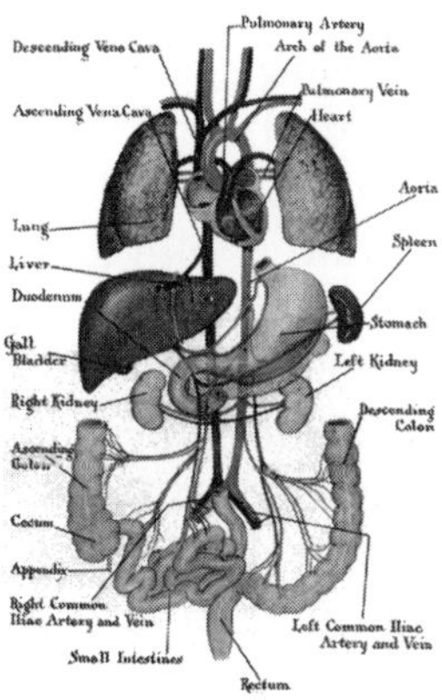

- **dihybrid cross**

  in genetics, a cross that involves two sets of characteristics.

- **dike**

  a discordant pluton that is substantially wider than it is thick. Dikes are often steeply inclined or nearly vertical. The expansion of a rock's volume caused by stress and deformation.

- **dinoflagellates**

  single-celled to colonial protistans characterised by two flagella, one girdling the cell and the other trailing the cell.

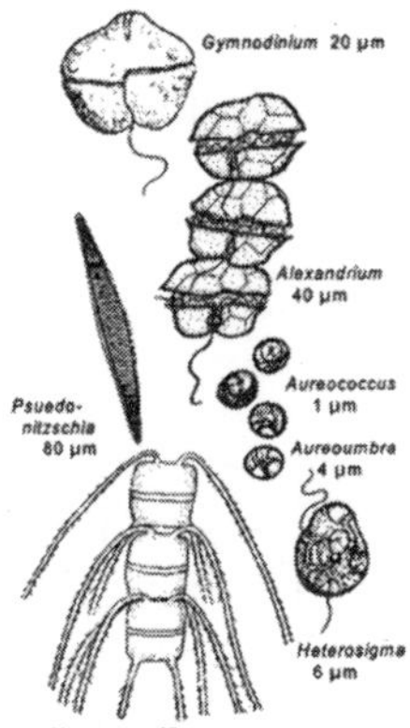

- **dinosaurs**

  any of the Mesozoic reptiles belonging to the groups designated as ornithischians and saurischians.

- **diorite**

  any of a group of dark, phaneritic, intrusive rocks that are the plutonic equivalents of andesite.

■ **dioxins**
chemical compounds, most are not toxic. The chlorinated dioxin (chlorinated hydrocarbon) may be very toxic.

■ **dip**
the angle formed by the inclined plane of a geological structure and the horizontal plane of the Earth's surface.

■ **diploid**
cells that contain homologous chromosomes. The number of chromosomes in the cells is the diploid number and is equal to 2n (n is the number of homologous pairs).

■ **dips**
economical, relativity trouble free structures for providing effective drainage of woods roads. Dips are considerably lower in cost than culverts, so time spent in careful construction is well justified.

■ **dip-slip fault**
a fault in which two sections of rock have moved apart vertically, parallel to the dip of the fault plane.

■ **direct or broadcast seeding**
1. sowing seed for broad coverage from the air or on the ground.
2. seeding of forest stands, roadways or specified plots for wildlife.

■ **direct seeding**
1. the application of tree seed to a denuded area to regenerate it with commercially valuable species.
2. spreading seed s over the forest seedbed by hand or machine.
3. practice used to assist or supplement natural seed fall and to achieve regeneration.

■ **directed pressure**
force exerted on a rock along one plane, flattening the rock in that plane and lengthening it in the perpendicular plane.

■ **directional felling**
1. predetermining the way a tree will land when it hits the ground. When shears are used, the wedge-shaped

blade provides a lever that directs the tree into its lay.
2. felling trees according to a predetermined pattern or lay on the ground. It may involve jacking or pulling of trees.

- **directional selection**
a process of natural selection that tends to favour phenotypes at one extreme of the phenotypic range.

- **dirty chips**
wood chips which include bark and some leaves or needles, as would be used in a fuel or biomass operation. No debarker step required before chipping.

- **disaccharides**
1. sugars made up of two monosaccharides held together by a covalent bond; e.g., sucrose and lactose.
2. type of sugar (saccharide) composed of two sugar molecules bonded together with an ester (covalent) bond examples include sucrose, maltose and lactose.

- **disappearing stream**
a surface stream that drains rapidly and completely into a sinkhole.

- **disc trencher**
a machine designed for mechanical site preparation. It provides continuous rows of planting spots rather than intermittent patches as provided by patch scarifiers. It consists of scarifying steel discs equipped with teeth.

- **discontinuous variation**
occurs when the phenotypes of traits controlled by a single gene can be sorted into two distinct phenotypic classes.

- **discretionary authority**
the power to make a decision where the choice of whether to make a decision is that of the decision maker.

- **disking (harrowing)**
a site preparation system where a heavy harrow with large disks is pulled over a site in order to eliminate selected vegetation.

- **dispersed retention**
retaining individual trees scattered throughout a cutblock.

- **displaced terrane**
a fault-bounded body of rock - sometimes thousands of square kilometres

in area - that originated elsewhere geographically and has then moved, perhaps long distances, by plate motion.

- **disruptive selection**
  a process of natural selection that favours individuals at both extremes of a phenotypic range.

- **dissolution**
  a form of chemical weathering in which water molecules, sometimes in combination with acid or another compound in the environment, attract and remove oppositely charged ions or ion groups from a mineral or rock.

- **dissolved load**
  a body of sediment carried by a stream in the form of ions that have dissolved in the water.

- **dissolved oxygen**
  a measure of the amount of oxygen dissolved in water. Expressed as mg/l of O2.

- **distal tubule**
  the section of the renal tubule where tubular secretion occurs.

- **distal**
  away from (contrast with proximal).

- **distributary**
  one of a network of small streams carrying water and sediment from a trunk stream into an ocean.

- **disturbance**
  1. a discrete event, either natural or human-induced, that causes a change in the existing condition of an ecological system.
  2. a natural or human-induced environmental change that alters one or more of the floral, faunal and microbial communities within an ecosystem. Timber harvesting is the most common human disturbance. Windstorms and fire are examples of natural disturbance.

- **ditch block**
  a blockage that is located directly downgrade of a cross-drain culvert or cross-ditch and designed to deflect water flow from a ditch into a cross-drain culvert.

- **divergence**
  the process by which two lithospheric plates separated by rifting move farther

apart, with soft mantle rock rising between them and forming new oceanic lithosphere. See also **convergence**.

- **divergent evolution**
  the divergence of a single interbreeding population or species into two or more descendant species.

- **divergent plate boundary**
  the boundary between two tectonic plates that are moving apart.

- **diversity**
  quality, state, fact or instance of being different

- **DNA fingerprinting**
  a method to isolate and create images of DNA sequences. The image created is an identification aid for organisms similar to that of taking human fingerprints.

- **DNA hybridisation**
  the formation of hybrid DNA molecules that contain a strand of DNA from two different species. The number of complementary sequences in common in the two strands is an indication of the degree of relatedness of the species.

- **DNA ligase**
  in recombinant DNA technology, an enzyme that seals together two DNA fragments from different sources to form a recombinant DNA molecule.

- **DNA polymerase**
  in DNA replication, the enzyme that links the complementary nucleotides together to form the newly synthesised strand.

- **DNA sequencing**
  the process of determining the exact order of the bases in a DNA segment.

- **DOB (Diameter Outside Bark)**
  the diameter of a tree or log including bark thickness.

- **dogs**
  pointed teeth located between the chainsaw bar and motor. Used in falling or bucking to pivot saw and to maintain position while cutting.

- **dolostone**
  a sedimentary rock composed primarily of dolomite, a mineral made up of calcium, magnesium, carbon, and oxygen. Dolostone is thought to

form when magnesium ions replace some of the calcium ions in limestone, to which dolostone is similar in both appearance and chemical structure.

- **dome**

a round or oval bulge on the Earth's surface, containing the oldest section of rock in its raised, central part. The area from which water flows into a stream. Also called a watershed.

- **dominance**

the property of one of a pair of alleles that suppresses the expression of the other member of the pair in heterozygotes.

- **dominance hierarchy**

a social structure among a group of animals in which one is dominant and the others have subordinate nonbreeding positions.

- **dominant**

1. trees with crowns extending above the general level of the canopy and receiving full light from above and partly from the side; taller than the average trees in the stand with crowns well developed.
2. the overstory life form or species in a plant community which contributes the most cover or basal area to the community, compared to other life form or species.
3. refers to an allele of a gene that is always expressed in heterozygotes.

- **dominant trees**

trees or shrubs with crowns receiving full light from above and partly from the side; usually larger than the average trees or shrubs in the stand, with crowns that extend above the general level of the canopy and that are well developed but possibly somewhat crowded on the sides. A dominant tree is one which generally stands head and shoulders above all other trees in its vicinity. However, there may be a young, vigorous tree nearby, but not overtopped by a dominant tree. This smaller tree may be considerably shorter than the dominant, but still be receiving full light from above and partly from the sides. In its own immediate environment, it is dominant and should be recorded as such. Only understory trees immediately adjacent to the overstory tree will be assigned subordinate crown classes.

- **domino falling**
  placing undercuts and backcuts in a series of trees, then pushing them with another tree. Domino falling is an unsafe and dangerous practice.

- **dorsal**
  of or belonging to the upper surface; top.

- **dorsoventral**
  from top to bottom; from the upper surface to the lower surface.

- **dorsum**
  the upper surface; top.

- **dose**
  a measured concentration of a toxicant for a known duration of time (concentration per unit time).

- **dot grid**
  a transparent sheet of film (overlay) with systematically arranged dots, each dot representing a number of area units. Used to determine areas on maps, aerial photos, plans, etc.

- **double fertilisation**
  a characteristic of angiosperms in which a pollen tube carries two sperm cells to the female gametophyte in the ovule. One sperm cell fuses with the egg cell and gives rise to a diploid embryo. The other sperm cell fuses with the two polar cells to form a triploid cell that develops into the endosperm.

- **down log**
  portion of a tree that has fallen or been cut and left in the woods.

- **down-rated bridges**
  bridges whose carrying capacity has been reduced.

- **doyle rule**
  log rule used in the Eastern and Southern United States. Underestimates board foot-

age in small logs and overestimates in large logs.

- **drag scarification**
  a method of site preparation that disturbs the forest floor and prepares logged areas for regeneration. Often carried out by dragging chains or drums behind a skidder or tractor.

- **drainage basin**
  area of the earth's surface from which surface drainage all flows to a single outlet stream (a watershed in North America).

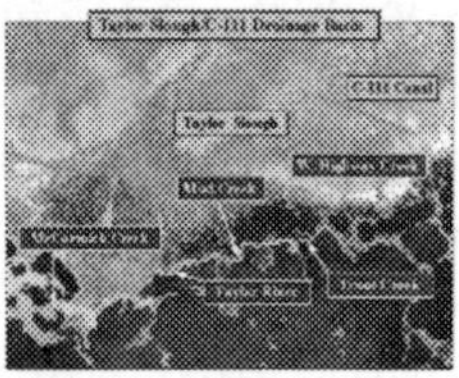

- **drainage divide**
  an area of raised, dry land separating two adjacent drainage basins.

- **drainage pattern**
  the arrangement in which a stream erodes the channels of its network of tributaries.

- **drainage structure**
  any device or land form constructed to intercept and/or aid in surface water drainage.

- **drainage system**
  a system designed to control the flow of water within a road prism.

- **drawbar**
  a fixed or hinged tow bar extending out the rear of a tractor.

- **drawdown**
  the process of reducing allowable annual cuts to a sustainable level.

- **dressing mill**
  where kiln-dried wood is planed smooth.

- **drop line**
  the line from a carriage to which chokers are set.

- **drum**
  a spool around which a cable is wound.

- **drum chopping**
  a site preparation technique in which logging debris is leveled by a bulldozer pulling a large drum filled with water. Chopped areas are often burned to further reduce debris and control sprouting before seedlings are planted.

- **drumlin**
  a long, spoon-shaped hill that develops when pressure

from an overriding glacier reshapes a moraine. Drumlins range in height from 5

to 50 meters and in length from 400 to 2000 meters. They slope down in the direction of the ice flow.

- **duff**
  1. the layer of partially and fully decomposed organic materials lying below the litter and immediately above the mineral soil. It corresponds to the fermentation and humus layers of the forest floor. When moss is present, the top of the duff is just below the green portion of the moss.
  2. the accumulation of needles, leaves and decaying matter on the woodland floor.

- **duff layer**
  the layer of loosely compacted debris underlying the litter layer on the forest floor.

- **dump**
  an area where logs are offloaded on land or into water.

- **dune**
  a usually asymmetrical mound or ridge of sand that has been transported and

deposited by wind. Dunes form in both arid and humid climates.

- **duodenum**
  the upper part of the small intestine.

- **duplication**
  an extra copy of a chromosome segment without altering the number of chromosomes.

- **dust palliatives**
  chemicals or compounds applied to road surfaces to reduce dust created by traffic.

- **dutchman**
  a flat area produced when two horizontal cuts of the undercut do not meet at a

point, thereby leaving a portion of an undercut not removed. An unsafe falling practice.

- **dynamothermal metamorphism**
  a form of regional metamorphism that acts on rocks caught between two converging plates and is initially caused by directed pressure from the plates, which causes some of the rocks to rise and others to sink, sometimes by tens of kilometres. The rocks that fall then experience further dynamothermal metamorphism, this time caused by heat from the Earth's interior and lithostatic pressure from overlying rocks.

- **early seral stage forest**
  stage of forest development that includes seedling, sapling, and pole-sized trees.

- **eco**
  this prefix come from the Greek 'Oikos' which means house. In the original context, ecology refers to the house we live in and economy refers to how we manage that house.

- **ecological balance**
  a state of dynamic equilibrium within a community of organisms in which genetic, species and ecosystem diversity remain relatively stable, subject to gradual changes through natural succession.

- **ecological classification**
  an approach to categorising and delineating, at different levels of resolution, areas of land and water having similar characteristic combinations of the physical environment (such as climate, geomorphic processes, geology, soil and hydrologic function), biological communities (plants, animals, microorganisms and potential natural communities) and the human dimension (such as social, economic, cultural and infrastructure).

- **ecological health**
  both the occurrence of certain attributes that are deemed to be present in a healthy, sustainable resource, and the absence of conditions that result from known stresses or problems affecting the resource.

▪ **ecological integrity**
the quality of a natural unmanaged or managed ecosystem in which the natural ecological processes are sustained, with genetic, species and ecosystem diversity assured for the future.

▪ **ecological reserve**
areas of Crown land which have the potential to satisfy one or more of the following criteria: areas suitable for scientific research and educational purposes associated with studies in productivity and other aspects of the natural environment; areas which are representative of natural ecosystems; areas in which rare or endangered native plants or animals may be preserved in their natural habitat; and areas that contain unique geological phenomena.

▪ **ecological status**
the degree of similarity between the present Community and the potential natural community on a site. Ecological status is rated irrespective of management objectives. Ratings are based on the floristic similarity of the current vegetation to the potential natural community. The similarity can be expressed on a relative scale ranging from 0 to 100 with adjective ratings assigned as low, moderate or high similarity.

▪ **ecological succession**
the changes, over time, in the structure and function of an ecosystem. When no previous vegetation exists on a site, the process is called primary succession. When a site supported vegetation previously but was disturbed, the process is called secondary succession.

▪ **ecological type (habitat type)**
a category of land having a unique combination of potential natural community; soil, landscape features, climate, and differing from other ecological types in its ability to produce vegetation and respond to management. Classes of ecological types include all sites that have this unique combination of components with the defined range of properties.

ecological unit
the map unit developed for an ecological type or types. This unit often includes a complex of small and intricately associated ecological types too small to delineiate separately.

ecologically suitable species
coniferous or deciduous tree species that are naturally adapted to a site's environmental conditions, including the variability in these conditions that may occur over time.

ecology
1. the science or study of the relationships between organisms and their environment.
2. the study of ecosystems. As a science ecology makes no value judgements.
3. the study of interactions between living organisms and their environment.

economic maturity
the point in the life cycle of a tree or stand when harvesting can be most profitable, i.e., when the rate of value increase of an individual tree or stand falls below a desired alternative rate of return.

economically operable
forest stands for which log prices exceed harvesting costs, including profit and return to capital.

economy
a system of producing, distributing and consuming wealth.

ecoregion classification
the ecoregion classification system is used to stratify B.C.'s terrestrial and marine ecosystem complexity into discrete geographical units at five different levels. The two highest levels, Ecodomains and Ecodivisions, are very broad and place B.C. globally. The three lowest levels, Ecoprovinces, Ecoregions and Ecosections, are progressively more detailed, narrow in scope and relate segments of the province to one another. They describe areas of similar climate, physiography, oceanography, hydrology, vegetation and wildlife potential.

ecosystem
1. a functional unit consisting of all the living organisms (plants, animals and mi-

crobes) in a given area, and all the non-living physical and chemical factors of their environment, linked together through nutrient cycling and energy flow. An ecosystem can be of any size-a log, pond, field, forest, or the earth's biosphere-but it always functions as a whole unit. Ecosystems are commonly described according to the major type of vegetation, for example, forest ecosystem, old-growth ecosystem, or range ecosystem.

2. a natural system which functions as a unit. It can be anything from a rotting log to the entire planet. It is assemblage of living organisms together with their non-living environment in a particular area.

3. a system made up of a community of animals, plants, and bacteria and its interrelated physical and chemical environment.

4. a natural unit comprised of living organisms and their interactions with their environment, including the circulation, transformation, and accumulation of energy and matter.

- **ecosystem integrity**
  the soundness or wholeness of the processes and organisms composing the ecosystem.

- **ecosystem management**
  the use of an ecological approach to achieve productive resource management by blending social, physical, economic and biological needs and values to provide healthy ecosystems.

- **ecosystem productivity**
  the ability of an ecosystem to produce, grow or yield products - whether trees, shrubs or other organisms.

- **ecosystem/cover type**
  the native vegetation ecological community considered together with non-living factors of the environment as a unit and, the general cover type occupying the greatest percent of the stand location. Based on tree or plant species forming a plurality of the stocking within the stand. May be observed in the field or computed from plot measurements.

- **ecotone**
  a transition area between two adjacent ecological

communities usually exhibiting competition between organisms common to both.

- **ecotype**

a genetic subdivision of a species resulting from the selective action of a particular environment and showing adaptation to that environment. Ecotypes may be geographic, climatic, elevational or soil related. Red maples and northern red oaks are both adapted to moist soils, but can also be found on drier sites where the genetic difference is their enhanced ability to retain water.

- **ectosymbiotic**

organism living on the surface of another organism.

- **edatope**

refers to a specific combination of soil moisture regime and soil nutrient regime.

- **edge**

1. the outer band of a patch that has an environment significantly different from the interior of the patch.
2. the transition between two different types or ages of vegetation.
3. where plant communities meet or where successional stages or vegetative conditions with plant communities come together.
4. the boundary between open land and woodland or between any two distinct ecological communities. This transition area between environments provides valuable wildlife habitat for some species, but can be problematic for sensitive species, due to increased predation and parasitism.

- **edge effect**

1. habitat conditions (such as degree of humidity and exposure to light or wind) created at or near the more-or-less well-defined boundary between ecosystems, as, for example, between open areas and adjacent forest.
2. the drastically modified environmental conditions along the margins, or edges, of forest patches surrounded partially or entirely by harvested lands.

- **edge/area ratio**

length of forest edge per cutblock area.

- **education**

the process of training and developing the knowledge, skill, mind, character, etc., especially by formal schooling.

- **effective old-growth forest**
  old-growth forest largely unmodified by external environmental influences from nearby, younger forest stands.

- **effective rooting depth**
  the depth of the soil that accounts for 80 percent of the roots.

- **efficiency factor**
  a ratio corresponding to the average percentage of total time that productive work is being performed by personnel or equipment.

- **efficiency hour**
  identical to 'efficiency factor', but expressed in terms of time, such as a 50 minute hour (I.e.-50 minutes of productive time per 60 minute hour).

- **effluent**
  the liquid 'leftovers' that result from manufacturing pulp and paper, including chemical compounds and wood fibre fragments.

- **egg gallery**
  a long, narrow tunnel along the sides of which eggs are deposited in small niches; the pattern of construction is often diagnostic of a particular species of insect.

- **egg mass**
  cluster of eggs, usually in a matrix of body hairs or wing scales from the female adult and/or a mucilaginous cementing secretion.

- **ejectisome**
  a type of explosive extrusome found in cryptoflagellates.

- **electronic scanner**
  a light beam passes over an object to measure it and/or check information.

- **electrophoresis**
  a lab procedure in which DNA fragments are separated in a gel that is placed in an electric field. This is one of the steps in DNA fingerprinting.

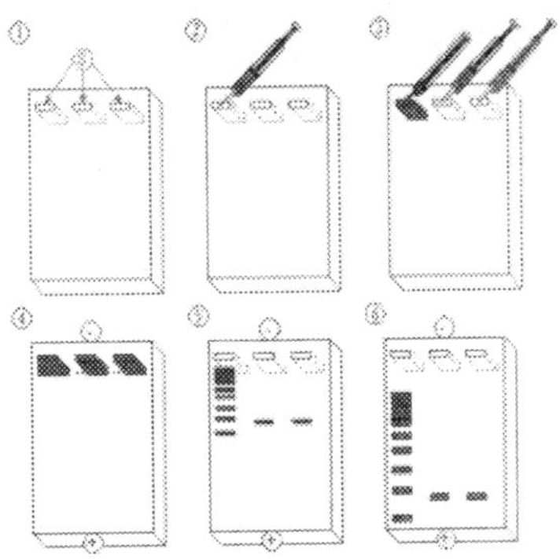

- **electroporation**
  one method of transformation in which a weak electric current creates pores in the cell membrane of a cell for entry of the new genes.
- **element**
  an identifiable component, process or condition of an ecosystem.
- **elemental chlorine**
  a pure form of chlorine gas which was used during pulp and paper bleaching processes before alternative methods were discovered.
- **elytra**
  the anterior leathery or chitinous wings of beetles and leafhoppers; serve as coverings to the hind wings and commonly meet at rest in a straight line down the middle of the dorsum.
- **embeddedness**
  a rating of the degree that larger substrate particles (boulder, rubble or gravel) are surrounded or covered by fine sediment.
- **emergent wetlands**
  a class of wetland dominated by grasses, sedges, rushes, forbs, and other rooted, water-loving (possibly broad-leaved) herbaceous plants that emerge from the water or soil surface; marshes are an example.
- **end haul**
  to move excavated material from one section of the road to another or to a disposal site, during road construction or modification.
- **endangered or threatened species**
  a species is endangered when the total number of remaining members may not be sufficient to reproduce enough offspring to ensure survival of the species. A threatened species exhibits declining or dangerously low populations but still has enough members to maintain or increase numbers.
- **endemic species**
  a species whose natural occurrence is confined to a certain region and whose distribution is relatively limited.
- **endurers**
  plants that re-sprout following a fire or endure the effects of fire.

- **engineered wood products**
  a composite wood product using glued fibre, lumber and/or veneer to meet specific design criteria.

- **engraver**
  beetle which feeds in the phloem-cambium region of woody plants, often scoring or engraving adjacent sapwood tissues.

- **enhancement**
  emphasis on improving the value of particular aspects of water and related land resources.

- **entrainment**
  mobilisation, by flowing water, of sediment or organic debris from the bed or banks of a stream channel.

- **entrenched**
  a legislative requirement which previously may only have been required by contract or policy.

- **entrepreneur**
  a person who organises and manages a business undertaking, assuming the risk for the sake of the profit

- **environment**
  1. the interaction of climate, soil, topography, and other plants and animals in any given area. An organism's environment influences its form, behavior and survival. 2. the sum of all external conditions and influences, living and nonliving, that effect the development and, ultimately, the survival of an organism or group of organisms.

- **environmental assessment**
  a process designed to contribute pertinent environmental information to the decision-making process of forest management and other resource projects and programs.

- **environmental rehabilitation**
  measures undertaken to remedy environmental damage done to the land.

- **Environmentally Sensitive Areas (ESAS)**
  areas requiring special management attention to protect important scenic values, fish and wildlife resources, historical and cultural values and other natural systems or processes. ESAs for forestry include potentially fragile, unstable soils that may deteriorate

unacceptably after forest harvesting, and areas of high value to non-timber resources such as fisheries, wildlife, water and recreation.

- **enzyme**

1. a protein that acts as a catalyst, accelerating the rate of specific biochemical reactions.

2. chemicals that break starch (complex carbohydrates) down to usable glucose. Used by insects and fungi to break down wood.

- **ephemeral stream**

a stream that flows only during or shortly after a storm event.

- **epicormic branching**

branches arising from buds in bark along mainstem, most commonly occurring in trees under crown stress; also called watersprouts.

- **epidemic**

a change, usually a sudden increase, in a disease within a population.

- **epithelium**

tissue covering or lining of the surface of an organism or an organ, both external and internal. Examples include the skin cells, the inner linings of the lungs, blood vessels, and the digestive tract, etc. Epithelial cells are generally tightly packed, providing a continuous barrier protecting the underlying cells.

- **equalising block**

a block used to distribute load as in connecting anchor cables to heel lines in a multiple anchor system.

- **Equivalent Basal Diameter (EBD)**

the square root of sum of squared basal diameters (BD).

- **permanent seasonal roads**

these are maintained as part of the permanent road system but are designed for use only when the ground is frozen or firm. These roads are generally narrower than permanent all-season roads, are built to lower engineering standards and have minimal surface gravel.

- **erosion**

1. the wearing away of the land surface by rain, run-

ning water, wind, ice, gravity or other natural or anthropogenic agents.

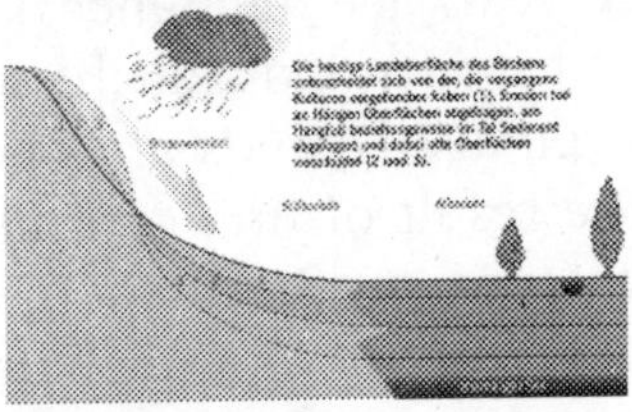

2. the process by which the surface of the earth is worn away by the action of wind or water in the form of rain drops, surface runoff and waves.

- **erosion severity**

the degree of erosion taken place on the site. Classes include:

Slightly eroded: Erosion has changed the soil enough to require only slight modification of management from that of the non-eroded soil; potential use and management remain generally the same. Most slightly eroded soils have class 1 erosion. Slightly eroded areas are not distinguished from non-eroded areas in most surveys.

Moderately eroded: Erosion has changed the soil to such an extent that required management or the response to management differs in major respects from that of the non-eroded soil. Suitabilities for major uses such as range or forestry, and major engineering uses—are the same. Shallow gullies may be present. The word 'moderately' is omitted from the name unless it is needed to differentiate between this phase and other eroded phases of the same soil.

Severely eroded: Erosion has changed the soil so much that

1. the eroded soil is suited only to uses significantly less intensive than the non-eroded soil,
2. the eroded soil needs intensive management immediately or over a long period to be suitable for the same uses as the non-eroded soil,
3. productivity is reduced significantly, or
4. limitations for some major engineering interpretations are greater than on the non-eroded soil. Shallow gullies, or a few deep ones, are common in some places.

- **escape route**
a planned and brushed out route used by fallers to make their way into the clear when the backcut is completed and the tree begins to fall.

- **established seedlings**
tree seedlings which are expected to survive and continue to grow over the next several years.

- **estuary**
a coastal body of water that has a connection with salt and fresh water sources.

- **evaporation**
the process of changing from a liquid to a vapour.

- **evapotranspiration**
the evaporation of water from the soil and the transpiration of water from the plants that live in that soil. Approximately one-quarter of a forest's annual rainfall returns to the air through evapotranspiration.

- **even flow**
in harvest scheduling, the requirement that the harvest level in each period be equal to the harvest level in the preceding period.

- **even-aged**
a forest stand or forest type in which relatively small (10-20 year) age differences exist between individual trees. Even-aged stands are often the result of fire, or a harvesting method such as clearcutting or the shelter wood method.

- **even-aged management**
1. forest management method in which all trees in an area are harvested at one time or in several cuttings over a short time to produce stands that are all the same age or nearly so. This management method is commonly applied to shade-intolerant conifers and hardwoods.
2. the silvicultural management of a stand where all trees in an area are harvested to produce a new stand that consists of one age class of trees. Even-aged management practices include clearcutting, seed-tree and shelter wood methods.

- **even-aged silvicultural system**
a silvicultural system that is designed to regenerate and maintain an even-aged

stand. Clearcutting, seed tree, and shelter wood are even-aged systems.

- **even-aged stand**
  a stand of trees consisting of one or two age classes. Even-aged stands are often the result of fire, or a harvesting method such as clearcutting or shelter wood.

- **evergreen**
  1. never entirely without green foliage, leaves persisting until a new set has appeared.
  2. trees with needles or leaves that remain alive and on the tree through the winter and into the next growing season.

- **excavated trail**
  a constructed trail that has a width greater than 1.5 m and a mineral soil cutbank height greater than 30 cm.

- **excavator**
  a hydraulically operated, boom equipped digging machine, most commonly track mounted, which can build full-benched logging roads with little to no sidecast loss.

- **existing visual condition**
  the present level of landscape alteration caused by resource development activities and expressed in terms of the visual quality objective categories.

- **exoskeleton**
  a supportive structure lying outside the cell or body.

- **exotic species**
  a species that is not native to the ecosystem; also known as invader species.

- **expansion**
  growing new woods and forests.

- **export log**
  a log that meets the species, ring count per inch, diameter, taper and knot size requirements for export to countries such as Japan.

- **extension**
  a line added to another to increase its length.

- **extension services**
  assistance provided to woodland operators. May include help with the preparation of forest management plans, cutting permits, marking trees for selective cutting, and guidance in carrying out slash disposal, site preparation, planting, etc.

■ **External Yarding Distance (EYD)**
the slope distance from the landing to the furthest reachable point within the cutting unit's boundary.

■ **extirpation**
the eradication of a species from a portion of its natural range.

■ **extracellular**
outside the cell.

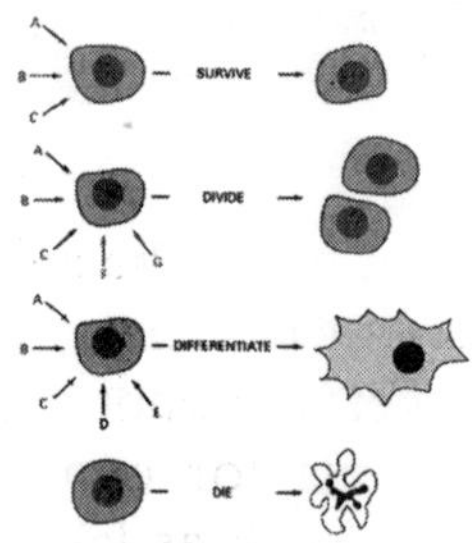

■ **extractives**
chemical 'wastes' which are stored in the heartwood of trees providing natural durability.

■ **Extra-Improved Plow Steel (EIPS)**
a standard grade of wire rope with the highest breaking strength, 15% higher than Improved Plow Steel (IPS) grade.

■ **extrude**
to push out.

■ **extrusome**
a kind of organelle, the contents of which can be extruded, for example to catch or kill prey or for protection.

■ **extrusopodium**
pseudopodia which bear extrusomes.

■ **eye**
a loop at the end of a wire rope secured by splicing or press fitted.

■ **factor of safety**
the ratio of the breaking strength of a line or material to the applied load.

■ **fairlead**
a device that consists of pulleys or rollers arranged to permit reeling in a cable from any direction; commonly found atop a standing tower yarder.

■ **falldown effect**
a decline in timber supply or harvest level associated with the transition from harvesting the original stock of natural mature timber over one rotation to harvesting at a non declining level (typically equal to the annual increment) after con-

version to a forest with a balanced age class structure.

- **faller**

a logger who fells timber; usually restricted to people doing hand felling, not to operators of tree harvesters. AKA 'Chopper', 'Feller'.

- **falling wedge**

wedge used to throw a tree in the desired direction.

- **fan-shaped setting**

a logging unit with yarding roads radiating from a common landing or common tailhold. Most commonly seen in cable units.

- **fasteners**

the hardware (e.g. nails, screws, bolts, joist hangers) used to secure treated wood. Since treated lumber is used for durability, fasteners should be hot-dipped galvanised or stainless steel, especially with water borne preservatives, which contain corrosive salt.

- **favourable**

descending grade in the direction of travel. Opposite of adverse.

- **fell**

to cut down trees.

- **feller-buncher**

1. a harvesting machine that cuts a tree with shears or a saw and then piles it.

2. a machine which fells trees using a mechanical shear or a disc saw as an attachment. A Feller-Buncher may accumulate several trees before creating just the right size bunch for a grapple skidder to take to the landing. A mechanical delimber or a whole tree chipper might be waiting at the landing to further process the tree.

- **feller-processor**

a machine that combines felling with one or more other functions, typically delimbing and bucking.

- **felling**

the process of cutting down standing trees.

- **felling and bucking**

the process of cutting down standing timber and then cutting it into specific lengths for yarding and hauling.

- **felling licence**

legal permission needed to fell trees.

■ **ferrule**

a metal sleeve or collar swaged to the end of a wire rope to make a terminal knob.

■ **fertilisation**

the addition of fertiliser to promote tree growth on sites deficient in one or more soil nutrients. Also used to improve the vigour of crop trees following juvenile spacing or commercial thinning.

■ **fibre**

in papermaking this refers to the cellulose fibre. Softwood trees produce long cellulose fibres that are good for strength and hardwood trees produce short fibres that are good for smoothness. Fibres are often blended to create the ideal characteristics for a given paper.

■ **fibre core**

the central member of a wire rope, made of woven strands of synthetic or natural fibres such as sisal, jute or cotton.

■ **filament**

1. a thin strand. May refer to the appearance of an organism or a strand of cytoplasm. In angiosperms: the slender, thread-like stalks that make up the stamens of a flower.
2. thread-like structures that support the anthers out from the flower base.

■ **fill**

material used to raise the desired road profile above the natural ground line.

■ **fill bank**

the fill material used to shape a road from the outer edge of the travelled portion to its intersection with the existing ground profile.

■ **fill slope**

1. the face of an embankment required to raise the desired road profile above the natural ground line.
2. the surface formed where earth is deposited to build a road or trail.

■ **fill-in planting**

planting required to supplement poorly stocked natural regeneration or to replace seedlings that have died on previously planted sites.

■ **filose pseudopodium**

pseudopodia that are thin and threadlike but without internal skeletal elements.

- **filter strip**

an area of land adjacent to a water body which acts to trap and filter out suspended sediment and chemicals before entering surface waters. Harvesting and other woodland management activities are permitted in the filter strip as long as the integrity of the filter strip is maintained and mineral soil exposure is kept to a minimum.

- **fine filter approach**

an approach to maintaining biodiversity that is directed toward particular habitats or individual species that might fall through the coarse filter. These habitats may be critical in some way and the species threatened or endangered.

- **fine paper**

1. paper used in making copy paper, computer forms, tablets, envelopes, printing paper, etc.
2. an assessment of both fixed and variable factors of the fire environment, which determine the ease of ignition, rate of spread, difficulty of control and the fire impact.
3. measure of the likelihood of a forest fire, based on temperature, relative humidity, wind force and direction, and
4. the dryness of the woods.

- **fire hazard**

1. the potential fire behavior for a fuel type, regardless of the fuel type's weather-influenced fuel moisture content or its resistance to fireguard construction. Assessment is based on physical fuel characteristics, such as fuel arrangement, fuel load, condition of herbaceous vegetation, and presence of elevated fuels.
2. condition of fuel on the ground, particularly slash.

- **fire impact**

the immediately evident effect of fire on the ecosystem in terms of biophysical alterations (e.g., crown scorch, mineral soil erosion, depth of burn, fuel consumption).

- **fire line**

cleared area extending down to mineral soil that surrounds a fire to prevent the fire from reaching fresh fuels.

- **fire management**
  the activities concerned with the protection of people, property and forest areas from wildfire and the use of prescribed burning for the attainment of forest management and other land use objectives, all conducted in a manner that considers environmental, social and economic criteria.

- **fire regime**
  the combination of fire frequency, predictability, intensity, seasonality and extent of fire in an ecosystem.

- **fire retardant**
  a substance that by chemical or physical action reduces flammability of combustibles.

- **fire risk**
  the probability or chance of fire starting determined by the presence and activities of causative agents.

- **fire season**
  the period(s) of the year during which firs are likely to start, spread and do damage to values-at-risk sufficient to warrant organised fire suppression; a period of the year set out and commonly referred to in fire prevention legislation. In B.C. the fire season is considered to extend from April 1 to October 31.

- **fire suppressant**
  an agent directly applied to burning fuels to extinguish the flaming and smoldering or glowing stages of combustion.

- **fire suppression**
  all activities concerned with controlling and extinguishing a fire following its detection. Synonymous with fire control.

- **fire trail**
  cleared area constructed around logging slash or other fire hazards in order to prevent the spread of fire to this hazardous material.

- **fire triangle**
  the three components (heat, fuel, and oxygen) necessary for a fire to take place and for the fire to keep burning.

- **fire watch**
  a look-out posted after the logging shift to watch for any fires starting from the day's activities.

- **firebreak**

1. any nonflammable barrier used to slow or stop fires. Several types of firebreaks are mineral soil barriers; barriers of green, slow-burning vegetation; and mechanically cleared areas. 2. a strip of land or trail/pathway that is maintained as an open corridor that can help slow down or stop wildfires. Trails, roads, rivers, canals, or utility rights-of-way may be considered as firebreaks in times of an emergency if necessary.

- **fireguard**

a strategically planned barrier, either manually or mechanically constructed, intended to stop a fire or retard its rate of spread and from which suppression action is carried out to control a fire; the constructed portion of a control line.

- **fireline**

1. that portion of the fire upon which resources are deployed and actively engaged in suppression action. In a general sense, the working area around a fire. 2. a barrier used to stop the spread of fire, constructed by removing fuel or rendering fuel unflammable by use of water or fire retardants.

- **firmwood**

sound or solid wood in a log suitable for either chips or solid wood products such as lumber or veneer. Burnt wood, voids, and soft rots are not considered firmwood.

- **first order stream**

stream originating in a seepage zone or spring, with no entering tributaries; the most headward channels in the drainage network.

- **first pass**

the first of two or more planned entries over a specific period of time (usually one rotation) to harvest timber.

- **fish harvest**

pounds of fish harvested by commercial and subsistence anglers, and number of fish user days spent on recreational fishing.

- **fish-bearing waters**

lakes, streams, and ponds that have resident fish populations.

- **fisheries classification**

water bodies and streams classed as either having a cold water or warm water fishery. This designation is dependent upon the dominate species of fish occupying the water.

Warm water fisheries: Warm water fisheries support fish able to tolerate water temperatures above 80 degrees F. Warm water fish include such species as crappies, small and largemouth bass, sunfish, yellow perch, and catfish.

Cold water fisheries: Cold water fisheries support fish that prefer clear, cold waters; are not tolerant of extreme temperature changes; and cannot survive for long periods with temperatures above 68 degrees F. Species such as trout, salmon, and grayling are recognised as cold water fish.

- **fisheries stream class**

streams or portions of streams that are frequented by anadromous salmonids and/or resident game fish or regionally significant fish species; or streams that have been identified for fishery enhancement in an approved fishery management plan.

- **fisheries-sensitive zones**

side and back channels, valley wall ponds, swamps, seasonally flooded depressions, lake littoral zones and estuaries that are seasonally occupied by overwintering anadramous fish.

- **fishery**

one or more stocks of fish which can be treated as a unit for purposes of conservation and management, plus any fishing for such stocks.

- **fixation**

the chemical process in which the preservative metals in water-borne solution react with and bond to wood fibre molecules.

- **fixed area plot sampling method**

a controlled cruise method where small plots of a fixed size are used to sample a portion of a forest area to obtain information (such as tree volume) that can be used to describe the whole area.

- **fixed cost**

  a cost item for a piece of equipment which does not vary in the aggregate or total. In other words, the production or utilisation of the equipment does not change the total cost. e.g. interest, insurance, annual depreciation. However, the per unit cost may vary inversely as utilisation increases.

- **fixed-grip harvesting head**

  the harvesting head type found on feller-bunchers. Usually fixed to the boom or front of the harvester and does not rotate around all three axes.

- **flagellate**

  a kind of protist bearing flagella; a very diverse group with unclear boundaries. distinguished from ciliates because flagella are few in number, and usually create a thrust along the length of the organelle rather than parallel to the body surface.

- **flashboard risers**

  a system of removable boards or panels that can be installed on culvert outlets or in drainage ditches to regulate the amount and/or flow of water in ditches and the nearby ground water levels.

- **flat or straight planting**

  planting trees directly into the ground without bedding, or in some cases, without first removing logging debris.

- **fleck**

  white to tan necrotic lesions up to a few millimeters in length or diameter, usually confined to the upper surface of leaves.

- **flexible**

  adjustable to change; capable of modification.

- **flood discharge criteria**

  the volume of flood that a bridge or culvert must be designed to accommodate.

- **floodplain**

  a level, low-lying area adjacent to streams that is periodically flooded by stream water. It includes lands at the same elevation as areas with evidence of moving water, such as active or inactive flood channels, recent fluvial soils, sediment on the ground surface or in tree bark, rafted debris and tree scarring.

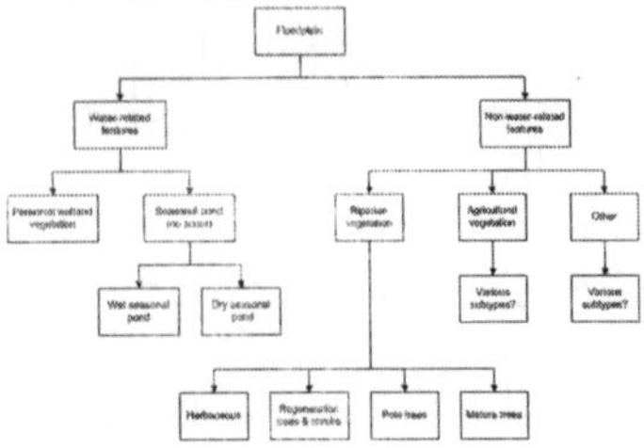

- **flute**
  one of the wave shapes pressed into corrugated medium. These are categorised by the size of the wave.

- **fluvial processes**
  all processes and events by which the configuration of a stream channel is changed; especially processes by which sediment is transferred along the stream channel by the force of flowing water.

- **fly**
  to carry logs completely off of the ground on a skyline.

- **flyrock**
  rock displaced by blasting and propelled beyond recoverable limits.

- **foliar analysis**
  chemical evaluation of the status of plant nutrients or the plant-nutrient requirements of a soil by the analysis of leaves or needles.

- **food chain**
  the transfer of food energy from organisms in one nutritional level to those in another.

- **forage**
  1. asses, herbs and small shrubs that can be used as feed for livestock or wildlife. 2. vegetation such as leaves, stems, buds and some types of bark, that can be eaten by animals.

- **forage utilisation**
  the proportion of current year's forage production that is consumed or destroyed by grazing animals. Forage is all browse and herbage that is available and acceptable to grazing animals.

- **forb**
  herbaceous plant other than those in the Gramineae (true grasses), Cyperaceae (sedges), and Juncaceae (rushes) families, i.e., any nongrass-like plant having little or no woody material.

- **ford**
  1. a dip constructed in the roadbed at a stream crossing, instead of a culvert or

bridge. The streambed must be of erosion-resistant material, or such material must be placed in contact with the streambed.

2. submerged stream crossing where the streambed may need to be reinforced to bear intended traffic. A place where a perennial or intermittent stream may be crossed by vehicle.

- **forest**

1. area managed for the production of timber and other forest products or maintained as wood vegetation for such indirect benefits as protection of catchment areas or recreation.

2. an ecosystem dominated by trees, with a unique combination of plants, animals, microbes, soil and climate.

3. a biological community dominated by trees and other woody plants.

- **forest canopy**

the cover of branches and foliage formed collectively by the crowns of adjacent trees and other woody growth.

- **forest cover**

forest stands or cover types consisting of a plant community made up of trees and other woody vegetation, growing more or less closely together.

- **forest cover map**

a map showing relatively homogeneous forest stands or cover types, produced from the interpretation of aerial photos and information collected in field surveys. Commonly includes information on species, age class, height class, site and stocking level.

- **forest cover type**

a descriptive term used to group stands of similar characteristics and species composition (due to given ecological factors) by which they may be differentiated from other groups of stands.

- **forest development plan**

an operational plan guided by the principles of integrated resource management (the consideration of timber and non timber values), which details the logistics of timber development over a period of usually five years. Methods, schedules, and responsibilities for accessing, harvesting, renewing, and protecting the resources are set out to enable site-specific operations to proceed.

- **forest ecology**
  the relationships between forest organisms and their environment.

- **Forest Ecosystem Network (FEN)**
  a planned landscape zone that serves to maintain or restore the natural connectivity within a landscape unit. A forest ecosystem network consists of a variety of fully protected areas, sensitive areas, classified areas and old- growth management areas.

- **forest fire**
  any wildfire or prescribed fire that is burning in forest, grass, alpine or tundra vegetation types.

- **forest floor**
  layers of fresh leaf and needle litter, moderately decomposed organic matter, and humus or well-decomposed organic residue.

- **forest floor (litter) and humus**
  the freshly cast, partly decomposed and fully decomposed vegetative material on the soil surface.

- **forest floor displacement hazard**
  a ranking of the potential adverse impacts on forest productivity resulting from removal of the accumulated organic matter that constitutes the forest floor.

- **forest fragmentation**
  the change in the forest landscape, from extensive and continuous forests of old-growth to mosaic of younger stand conditions.

- **forest health**
  a forest condition that is naturally resilient to damage; characterised by biodiversity, it contains sustained habitat for timber, fish, wildlife, humans and meets present and future resource management objectives.

- **forest health agents**
  biotic and abiotic influences on the forest that are usually a naturally occurring component of forest ecosystems. Biotic influences include fungi, insects, plants, animals, bacteria and nema-

todes. Abiotic influences include frost, snow, fire, wind, sun, drought, nutrients and human-caused injury.

- **forest health treatments**
  the application of techniques to influence pest or beneficial organism populations, mitigate damage, or reduce the risk of future damage to forest stands. Treatments can be either proactive (for example, spacing trees to reduce risk of attack by bark beetles) or reactive (for example, spraying insecticides to treat outbreaks of gypsy moth).

- **forest interior conditions**
  conditions found deep within forests, away from the effect of open areas. Forest interior conditions include particular microclimates found within large forested areas.

- **forest interior dependent species**
  animal species that depend upon extensive areas of continuous, unbroken forest habitat to live and reproduce, and are susceptible to higher rates of predation and population decline when interior forest habitat is fragmented or disturbed.

- **forest inventory**
  1. an assessment of forest resources, including digitised maps and a database which describes the location and nature of forest cover (including tree size, age, volume and species composition) as well as a description of other forest values such as soils, vegetation and wildlife features.
  2. land at least 10 percent stocked by trees of any size or formerly having had such tree cover and not currently built-up or developed for agricultural use. Forestland may include Grassland, Shrubland, Treeland, Wetland, and/or Barren land. Examples of forest land use are grazing, recreation and timber production.

- **forest land**
  1. land that is now, or is capable of becoming, at least 10 percent stocked with forest trees and that has not been developed for nontimber use.
  2. land which has as its highest and best use the growing and harvesting of trees,

including land which is being managed in accordance with a forest management plan approved under regulations, but does not include a farm.

- **forest land class**

a classification of an area based upon its capability of producing industrial wood (i.e., all commercial roundwood products except fuelwood), its legal status concerning timber utilisation, and its proximity to urban and rural development. Classes include:

Timberlan: Forest land that is producing or capable of producing in excess of 20 cubic feet per acre per year of industrial wood crops under natural conditions, that is not withdrawn from timber utilisation, and is not associated with urban or rural development. Currently inaccessible and inoperable areas are included.

Reserved Timberland: Public forest land, not associated with urban or rural development, that produces or is capable of producing in excess of 20 cubic feet per acre of industrial wood crops under natural conditions, on which statutory or administrative restrictions prohibit the harvest of trees.

Other Forest Land: Forest land not capable of producing 20 cubic feet per acre per year of industrial wood crops under natural conditions and not associated with urban or rural development. These sites often contain tree species that are not currently utilised for industrial wood production or trees of poor form, small size, or inferior quality that are unfit for industrial products. Unproductivity may be the result of adverse site conditions such as sterile soil, dry climate, poor drainage, high elevation, and rockiness. This land may or may not be withdrawn from timber utilisation.

Urban Forest Land: Forest land closely associated with or in such proximity to urban nonforest land uses that is not likely to be managed for the production of industrial wood products on a continuing basis. Wood removed would be for land clearing, fuelwood, or aesthetic purposes. Such forest land may be associated with industrial, commercial, residential, or recreational nonforest uses. The ecological character of the forest remains intact, i.e., the understory has not been removed or altered enough to preclude forest succession or replanting. Residential subdivisions, industrial parks, golf course perimeters, airport buffer strips, and public urban parks that qualify as forest land are included.

Developed Rural Forest Land: Forest land closely associated with or in such proximity to rural nonforest land uses that is not likely to be managed for the production of industrial wood products on a continuing basis. Wood removed would be for land clearing, fuelwood, or aesthetic purposes. Such forest land is usually associated with agricultural, recreational, or residential nonforest uses but may be associated with industrial or commercial uses as well. Examples of such nonforest uses include cropland, home sites, camping areas, and farmsteads. The ecological character of the forest remains intact, i.e., the understory has not been removed or altered enough to preclude forest succession or replanting.

- **forest landscape**

1. a portion of the land that the eye can see in one glance and in which the forest is the most dominant element.
2. land presently forested or formerly forested and not currently developed for nonforest use.

- **forest life cycle**

seeds fall on the soil, germinate and seedlings grow. When the trees become mature they produce seeds. They grow old, die and fall down. Fallen trees decompose, giving nutrients back to the soil - food for a new generation of seedlings.

- **forest management**

the practical application of scientific, economic and social principles to the administration and working of a forest for specified objectives. Particularly, that branch of forestry concerned with the overall administrative, economic, legal and social aspects and with the essentially scientific and technical aspects, especially silviculture, protection and forest regulation.

- **forest management cycle**
  the phases that occur in the management of a forest including harvesting, site preparation, reforestation, and stand tending.

- **forest management plan**
  a general plan for the management of a forest area, usually for a full rotation cycle, including the objectives, prescribed management activity and standards to be employed to achieve specified goals. Commonly supported with more detailed development plans.

- **forest mensuration**
  the measurement of volume, growth and development of individual trees and stands, and the various products obtained from them.

- **forest planning model**
  an analytical model (usually computer-based) that successively harvests and grows collections of forest stands over a period of several decades according to specific data and management assumptions.

- **forest practice**
  any activity that enhances and/or recovers forest growth or harvest yield, such as site preparation, planting, thinning, fertilisation and harvesting.
  Road construction or reconstruction within forest lands for the purpose of facilitating harvest or forest management. Any management of slash resulting from the harvest or improvement of tree species.

- **forest profile**
  the range of forest conditions that exists across the landscape, including such factors as timber species, quality, condition and age, location, elevation, topography, accessibility and economic viability.

- **forest recovery**
  the complex natural process by which floral, faunal and microbial communities respond to disturbance in the

forest ecosystem. More resilient ecosystems respond rapidly to disturbance, returning to the pre-disturbance ecological state within a relatively short time period (perhaps decades as opposed to centuries).

- **forest residuals**
  sum of wasted and unused wood in the forest, including logging residues; rough, rotten and dead trees; and annual mortality. Unmerchantable material normally left following conventional logging operations other than whole-tree harvesting.

- **forest tree breeding**
  the genetic study of trees to solve some specific problem or to produce a specially desired product.

- **forest tree improvement**
  the control of parentage combined with other silvicultural activities (such as site preparation or fertilising) to improve the overall yield and quality of products from forest lands.

- **forest type**
  groups of tree species commonly growing in the same stand because their environmental requirements are similar. North Carolina examples include pine and mixed hardwood; cypress, tupelo, and black gum; and oak and hickory.

- **forest type labels**
  the symbols which are used to code information about forest types on a forest cover map, such as site, disturbance, age and height class, species, stocking.

- **forest type lines**
  lines on a map or aerial photo outlining forest types.

- **forest types**
  associations of tree species that commonly occur because of similar ecological requirements. Massachusetts three major forest types are northern hardwoods, oak/hickory, white pine and oak/pine.

- **forest yield regulation**
  the administrative and technical process which facilitates yield control (regulation), often narrowly interpreted as a process that ensures regular and sustained forest yields.

- **forestation**

establishment of a forest, naturally or artificially, on an area, whether previously forested or not.

- **forested wetland**

an area characterised by woody vegetation over 20 feet tall where soil is at least periodically saturated with or covered by water.

- **forester**

a person engaged in the profession of forestry. In some countries the term is restricted to those who received formal post-secondary education in forestry or who possess the equivalent qualifications. A forester may or may not be a Registered Professional Forester, which is a legally-recognised title.

- **forestry**

1. the science, art and practice of managing and using for human benefit the natural resources that occur on and in association with forest lands.
2. the science of planting, managing and protecting forests.

- **Forestry Incentives Program (FIP)**

a federally funded cost-sharing program of the Agricultural Stabilisation and Conservation Service (ASCS). FIP provides payments to landowners who complete certain approved forest management practices, including site preparation, tree planting, and Timber Stand Improvement (TSI).

- **forewings**

front pair of wings.

- **forwarder**

a machine with a crane which can load logs onto its chassis and piggy-back them to a road where it can sort and pile them or load them directly onto a truck.

- **forwarding**

the act of carrying logs off the ground from stump to landing.

■ **fragmentation**
1. the process of transforming large continuous forest patches into one or more smaller patches surrounded by disturbed areas. This occurs naturally through such agents as fire, landslides, windthrow and insect attack. In managed forests timber harvesting and related activities have been the dominant disturbance agents.
2. the segmentation of a large tract or contiguous tracts of forest to smaller patches, often isolated from each other by nonforest habitat. Results from the collective impact of residential and commercial development, highway and utility construction, and other piecemeal land use changes.

■ **frass**
solid larval insect excrement; mixed with wood fragments in woodboring or bark-boring insects.

■ **free sheet paper**
paper made from cooked wood fibres mixed with chemicals and washed free of impurities. As compared to groundwood paper.

■ **free-growing assessment**
the determination for whether young trees have attained free- growing status.

■ **freshet**
high stream flow, usually confined to the stream channel and caused by a regularly recurring hydrological phenomenon (e.g., the snowmelt freshet) (regional term).

■ **fruit**
the seed-bearing product of a plant.

■ **fruiting body**
the reproductive part of a fungus that contains or bears spores. Also known as a conk.

■ **frustule**
the siliceous lorica of a diatom.

■ **fry**
1. the young stage of fishes (i.e., less than one year old), particularly after the yolk sac has been absorbed.

2. a juvenile salmonid that has absorbed its egg sac and is rearing in the stream.

- **fuel loading**
  a build-up of fuels, especially easily ignited, fast-burning fuels such as pine straw.

- **fuel management**
  the planned manipulation and/or reduction of living or dead forest fuels for forest management and other land use objectives (such as hazard reduction, silvicultural purposes, wildlife habitat improvement) by prescribed fire, mechanical, chemical or biological means and/or changing stand structure and species composition.

- **fuel model**
  mathematical descriptions of fuel properties (e.g. fuel load and fuel depth) that are used as inputs to calculations of fire danger indices and fire behaviour potential.

- **fuel moisture**
  the extent to which fuel will burn is largely determined by the amount of water in the fuel. Fuel moisture is a dynamic variable controlled by seasonal, daily and immediate weather changes. Fuel moisture, expressed as a percent, is computed from the weight of contained water in fuel divided by the oven dry weight of the fuel.

- **fuel break**
  an existing barrier or change in fuel type (to one that is less flammable than that surrounding it), or a wide strip of land on which the native vegetation has been modified or cleared, that act as a buffer to fire spread so that fires burning into them can be more readily controlled. Often selected or constructed to protect a high value area from fire.

- **fuel-loading**
  a build up of fuels, especially easily ignited, fast-burning fuels such as pine straw.

- **fuelwood**
  trees used for the production of firewood logs or other wood fuel.

- **full bench cut**
  forming the roadway entirely in cut.

- **full tree harvester**
  a harvesting machine that severs the tree and removes it whole from the forest

■ **full-tree harvesting**
a tree harvesting process that includes removing the trunk, branches and in some instances the roots from a forested site. In Canada this process is used to control root diseases.

■ **full-tree system**
a harvesting system in which the entire tree is hauled to the landing (usually by a skidder) after felling.

■ **fumigation**
the natural or controlled exposure of plants to toxic gases or volatile substances.

■ **fungi**
simple plant-like organisms that lack chlorophyll. They obtain their nutrition from living on or in other organisms (parasitically), from living with other organisms (symbiotically), or by breaking down dead organic materials (saprophytically). An example are the microscopic mycorrhiza which live off the tree, while fixing nitrogen for the tree. Mushrooms, molds, and yeast are included in this group.

■ **furans**
a colourless, liquid ring compound, prepared from wood tar or furfural, and used as a solvent for resins, plastics, etc., or as a tanning agent the fibrous materials from which paper is made.

■ **fusiform rust**
a disease resulting in a canker or swollen area on the limbs or trunks of pine trees from orange spores produced by infected oak leaves. Fusiform rust degrades stem quality and tree value, often leading to breakage, disfigurement and eventual death of the tree.

■ **gall**
a swelling or outgrowth of tissue induced by a pathogen or insect on a slant.

■ **gantry tower**
the short tower on a modern swing yarder, located

just to the rear of the leaning yarder tower, used to accomodate the walking guy lines and lines to raise and lower the boom of a yarding crane.

- **gap analysis**
  defining and analysing the gap between your current reality and your vision of the future. Your mission is to close this gap by selecting the best strategies.

- **gene expression**
  the production of a protein specified by a gene's sequence.

- **gene flow**
  the movement of genes among various populations of a species.

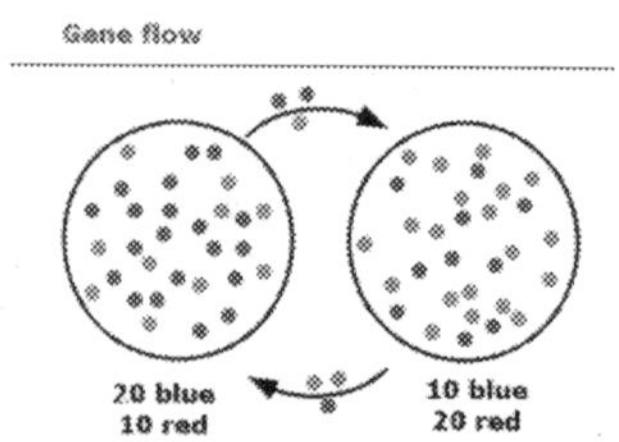

- **gene stacking**
  combining traits in a genetically modified living organism, for example, insect resistant and herbicide resistant genes.

- **gene**
  functional and physical unit of inheritance that is passed from parent to offspring. Genes are pieces of DNA containing the code for the synthesis of a functional polypeptide or RNA molecule. Most genes contain coding regions (exons), non-coding sequences (introns), and transcription-control regions.

- **genebank**
  a facility that stores plant germplasm of various species in the form of pollen, seeds, or in vitro culture.

- **generation**
  the successive developmental stages from reproduction to reproduction, e.g., egg, larva, pupa, adult.

- **genes**
  the tiny part of a cell that transfers the characteristics of the parent organism to its offspring

- **genetic code**
  a gene's 'manual' on how the cell makes specific proteins.

- **genetic diversity**
  1. variation among and within species that is attrib-

utable to differences in hereditary material.

2. the differences or variations in the genetic traits found in a species and among different species.

- **genetic engineering**
  a method used to transfer DNA from one organism into another that results in a genetically modified organism.

- **genetic map**
  locations of genes mapped along a chromosome marked with probes.

- **genetically improved seed and/ or vegetative propagules**
  seed or propagule that originate from a tree breeding program and that have been specifically designed to improve some attribute of seeds, seedlings or vegetative propagules selection.

- **genome**
  all the genes contained in an organism's set of chromosomes that direct its development. In animals, this includes the nuclear and mitochondrial DNA. In plants, this includes the nuclear, mitochondrial, and chlorop-last's DNA.

- **genotype**
  1. the entire genetic constitution, or the sum total of genes of an organism, in contrast to the phenotype.
  2. growth or development characteristics dependent on genetic information. The genetic constitution of an organism or a species in contrast to its observable characteristics.

- **Geographic Information System (GIS)**
  1. a computer system used for forestry forecasting, planning and monitoring.
  2. a computer system designed to allow users to collect, manage and analyse large volumes of spatially referenced information and associated attribute data.

■ **geologic features**
landforms or other features of significant geologic interest that may require special management to:
1. protect the special qualities, or
2. provide interpretation to the public.

■ **geologic formation**
a mappable body of rock identified by distinctive characteristics, some degree of internal homogeneity, and stratiographic position. The name normally consists of two parts. The first is the name of the geographic locality where the formation was first identified and described. This is followed by a descriptive geologic term, usually the dominant rock type.

■ **geologic hazards**
a natural condition that poses some risk to human health or safety.

■ **geologic time unit**
a division of time traditionally distinguished on the basis of observable changes in worldwide life forms as represented in the fossil record in sedimentary rocks.

■ **geotextile filter fabric**
a synthetic material placed on the flat, under road fill, with the primary functions of layer separation, aggregate confinement and distribution of load.

■ **germplasm**
set of genes that may be used for conservation purposes.

■ **gin pole**
erect pole with guys, 30 to 50 feet in height, leaning so

that a loading block hung at the top is directly over the point where the truck will be spotted for loading.

- **girdling**

1. to kill a tree by severing or damaging the cambium layer and interrupting the flow of food between the leaves and the rest of the tree. A method of 'brushing' carried out using a hatchet or special tool to cut through the bark and cambium of the tree.
2. a physical cutting or disruption of the cambial sap flow within a tree. Girdling by humans, animals, or insects can often kill a tree.
3. a method of killing unwanted trees by cutting through the living tissues around the bole. Can be used instead of cutting to prevent felling damage to nearby trees. Girdled trees can provide cavities and dead wood for wildlife and insects.

- **glacier**

a large mass of ice and snow that forms in areas where the rate of snowfall constantly exceeds the rate at which the snow melts.

- **glides**

a third general habitat category possessing attributes of both riffles and pools. Glides are characterised by moderately shallow water (10-30 cm deep) with an even flow that lacks pronounced turbulence. Although they are most frequently located at the transition between pool and the head of a riffle, glides are occasionally found in long, low gradient stream reached with stable banks and no major flow obstructions. The typical substrate is gravel and cobbles.

- **global warming**

increased levels of gases such as carbon dioxide that trap heat in the earth's lower atmosphere potentially causing global warming.

- **goal**

goals provide general purpose and direction. They are

the end result of ultimate accomplishment toward which an effort is directed. They generally should reflect perceived present and future need. They must be capable of being effectively pursued.

gout
formation of swellings at nodes or at the base of buds.

grabinski
a modified highlead yarding system where the butt rigging is attached to a riding block which rolls on the haulback line, providing limited deflection and suspension with a 2-drum yarder. Also nicknamed the 'Polish skyline'.

grade
the slope of a road or trail expressed as a percent of change in elevation per unit of distance travelled.

grade resistance (assistance)
the force of gravity resisting (assisting)movement on an adverse (favou rable) grade.

gradient
amount by which the grade increases or decreases in a unit of horizontal distance.

grading
classifying timber, lumber or logs according to quality or end-use.

grafting
a shoot or bud (scion) of one plant or tree is inserted into the stem or trunk of another, where it continues to grow, becoming a permanent part solid inclusions in cells or items adhering to the surface of cells. Usually refractile, in that they may look bright when viewed with the microscope.

grapple
a hinged set of jaws capable of being opened and closed, used to grip logs during yarding or loading. Can also

be attached to swing or nonswinging grapple skidder.

- **grapple skidder**
  a skidder equipped with a grapple to handle logs; used in place of chokers.

- **grapple yarding**
  cable yarding with grapples instead of chokers.

- **grass**
  a plant in the family Gramineae.

- **grass/forb**
  an early forest successional stage where grasses and forbs are the dominant vegetation.

- **grassland**
  areas on which vegetation is dominated by grasses, grass-like plants, forbs, and/or cryptogams (mosses. lichens, and ferns), provided these areas do not qualify as Built-up land or Cultivated cropland. Examples are tall grass and short grass prairies, meadows, cordgrass marshes, sphagnum moss areas, pasture lands and areas cut for hay.

- **grass-like plant**
  a plant which resembles grasses, but belongs to the families cyperaceae (sedges) and Juncaceae (rushes).

- **gravity logging**
  any cable system that depends on the force of gravity for the downhill travel of the carriage.

- **grazing lease**
  a lease of Crown land issued for grazing purposes under the Land Act.

- **grazing schedule**
  sets out the class and number of livestock that can use an area described in the schedule, the dates the livestock can use the area and other prescribed information.

- **grazing season**
  a period during which livestock may graze on Crown land under a grazing licence or grazing permit.

- **green manure**
  the nutrient rich, green foliage that can be returned to the site as a natural fertiliser.

- **green strip**
  uncut strip of timber left along streams and roads. Also known as buffer strip, leave strip, streamside management zone.

- **green tree retention**
  the reservation of live trees of a specific species and size from harvesting, to achieve a site-specific objective.

- **greenbelt**
  an extensive area of largely undeveloped or sparsely occupied land associated with a community set aside to contain development, preserve the character of the countryside and community and provide open space.

- **greened-up**
  a cutblock that supports a stand of trees that has attained the green-up height specified in a higher level plan for the area, or in the absence of a higher level plan for the area, has attained a height that is 3 m or greater, and if under a silvicultural prescription, meets the stocking requirements of that prescription, or if not under a silviculture prescription, meets the stocking specifications for that biogeoclimatic ecosystem classification specified by the regional manager.

- **gregarious**
  living in groups or communities.

- **gross total volume**
  volume of the main stem of the tree including stump and top. Volume of the stand including all trees.

- **ground lead**
  a method of, or condition in, cable logging in which the pull of the hauling line is parallel to the ground, i.e.-no deflection is provided and hence usually no to little lift.

- **ground skidding**
  pulling logs parallel to the ground without using an arch to raise the forward end.

- **ground truthing**
  the use of a ground survey to confirm the findings of an aerial survey or to calibrate quantitative aerial observations.

- **ground water**
  the subsurface water supply in the saturated zone below the level of the water table.

- **ground water aquifers**
  a geologic formation that is sufficiently permeable to conduct ground water and has the potential to yield economically significant quantities of water to wells and springs.

ground-based systems
logging systems that employ ground-based equipment such as feller-bunchers, hoe chuckers, skidders and forwarders.

groundwater
1. water below the level of the water table in the ground; water occupying the sub-surface saturated zone.
2. precipitation which infiltrates the ground and fills the pores in rocks and soil.

groundwood paper
newsprint and other inexpensive paper made from pulp created when wood chips are ground mechanically rather than refined chemically. Willamette does not make this type of paper.

grouser
the raised steel strip the length of a crawler pad to provide traction.

growing stock
the sum of all trees in a forest or specified part of it.

growth
increase in diameter, basal area, height, and volume of individual trees or stands during a given period of time. Also known as increment.

growth rate
the average annual percent change in population resulting from a surplus (or deficit) of births over deaths and the balance of migrants entering and leaving a country.

grub
an insect larva; a term loosely applied, usually to larvae of Coleoptera; larva is thick-bodied with well-developed thoracic legs but no abdominal prolegs.

grubbing and retention
removal of stumps, roots, embedded logs, organics and unsuitable soils before or concurrently with subgrade construction.

guild
species similar in their habitat needs as well as their response to habitat changes (e.g., ovenbird and woodthrush). One species in a guild is often used to represent the others when developing a stewardship management plan.

- **gully assessment procedure**
a procedure for determining gully sediment and debris transport potential, and suggested management strategies.

- **guy**
lines from a yarding machine to substantial anchors for support against yarding forces.

- **guyline**
a wire rope used as a guy.

- **gymnosperm**
any class of seed plants, mostly trees such as conifers, that produce naked seeds not enclosed in fruit.

- **habitat**
1. the place where an organism lives and/or the conditions of that environment including the soil, vegetation, water and food.
2. the area that provides an organism with adequate food, water, shelter, and living space, and/or the conditions of that environment including the soil, vegetation, water and food.
3. the region where a plant or animal naturally grows or lives; native environment.
4. the geographically defined area where environmental conditions (e.g. climate, topography, etc.) meet the life needs (e.g.,. food, shelter, etc.) of an organism, population or community.

- **habitat enhancement**
any manipulation of habitat that improves its value and ability to meet specified requirements of one or more species.

- **habitat management**
management of the forest to create environments which provide habitats (food, shelter) to meet the needs of particular organisms.

- **hack and squirt**
a method of conifer release and juvenile spacing where the bark of a tree is cut (hack) and herbicides are injected (squirt) to kill the tree.

- **hang up**
1. a log stuck behind a tree, stump, or other obstruction

during yarding, preventing its forward movement.
2. to fall a tree so that it lodges in a second tree rather than falling to the ground.
3. to get the saw bar caught in the kerf while bucking a log.

- **hardwood**

1. tree species with broad, flat, usually deciduous leaves, such as oak, maple, cottonwood and elm. Wood hardness varies a great deal and may actually be softer than some softwoods.
2. a deciduous or broadleaf tree; also applies to the wood from such trees.

- **hardwood trees**

trees that have broad leaves instead of needles.

- **harvest**

to remove trees from the forest for the purpose of using or selling them.

- **harvest cut**

the felling of the mature crop of trees either as a single clearcut or a series of regeneration cuttings.

- **harvest forecast**

the flow of potential timber harvests over time. A harvest forecast is usually a measure of the maximum timber supply that can be realised, over time, for a specified land base and set of management assumptions.

- **harvest pattern**

the spatial distribution of cutblocks and reserve areas across the forested landscape.

- **harvest schedule**

a document listing the stands to be harvested year or period, usually showing types and intensities of harvests for each stand, as well as a timetable for regenerating currently non-productive areas.

- **harvester**

a purpose-built machine with a head (attachment) for

felling and processing trees into log lengths. Sometimes the processing head itself is referred to as a Harvester.

- **harvesting**
  1. the practice of felling and removing trees or the removal of dead or damaged trees from an area.
  2. a loose term for the removal of trees from the forest, for product utilisation.

- **harvesting method**
  the mix of felling, bucking, and yarding (skidding) systems used in logging a stand of timber.

- **harvesting prescription**
  detailed plan on how, when, and where timber will be harvested from an area.

- **harvesting system**
  the mix of felling, bucking and yarding systems used in logging a stand of timber.

- **haulback block**
  a block in a cable yarding system used to guide the haulback line. In some systems, may also be called a tailblock.

- **haulback line**
  a wire rope used to pull the mainline with carriage or butt rigging with chokers back to the timber for the next turn.

- **hauling**
  a general term for the transportation of logs from one point to another, usually from a landing to the mill or shipping point.

- **hazard**
  a state that may result in an undesired event, the cause of risk. Hazard can apply to the probability of tree mortality or damage by an insect or disease and also represents material or fuel that will ignite and burn.

- **hazardous or danger tree**
  a tree or any component of a tree that has sufficient structural infirmity to be identified as having a high risk of falling and causing personal or property damage.

- **hazards, potential**
  a component of risk rating. Potential hazards are the

detrimental events that could result from inappropriate harvesting practices.

- **headrig**

principal machine in a sawmill, used for the initial breakdown of logs by sawing along the grain. Logs are first cut into cants on the head rig before being sent on to other saws for further processing. Also known as head saw.

- **healthy ecosystem**

an ecosystem in which structure and functions allow the maintenance of biodiversity, biotic integrity and ecological processes over time.

- **heartwood**

the inner core of a woody stem composed of nonliving cells and usually differentiated from the outer wood layer (sapwood) by its darker colour.

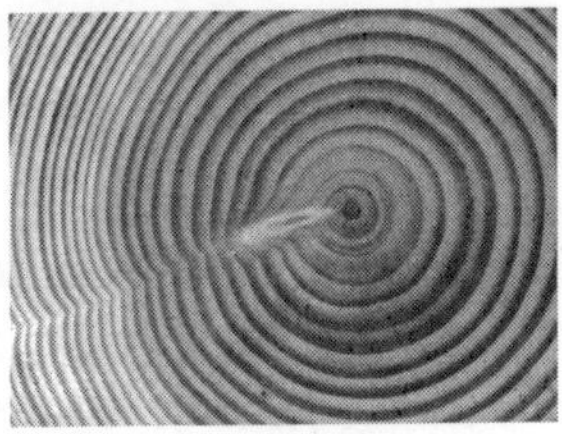

- **hectares**

approximately 2.5 acres.

- **heel boom**

a loader with a grapple or tongs at the end of a boom or arm which forces one end of the log being loaded against the underside of the boom to steady it and raise the front end of the log. Almost exclusively hydraulically operated today (but example of cable-operated heel booms can still be seen in some areas). AKA 'hydraulic loader' or a 'juicer'.

- **height class**

any interval into which the range of tree heights is divided for classification and use, commonly 3 m, 5 m, or 10 m classes.

- **height growth**

the increase in height over 5 years or the period between measurements (measured for coniferous trees).

- **height to crown, compacted**
  the vertical distance in feet from the ground to the base of the compacted live crown.
- **height to crown, uncompacted**
  the vertical distance in feet from the ground to the base of the live crown, measured to the lowest live branch-whorl or lowest live branch excluding epicormics.
- **height/diameter curve**
  a graphic representation of the relationship between individual tree heights and diameters used to determine tree volumes in localised areas.
- **helitack**
  initial attack on wildfires involving the use of helicopters and trained crews, deployed as a complete unit.

- **helitanker**
  a helicopter equipped with a helitank - a specially designed tank used for transporting and dropping suppressants or retardants.
- **helitorch**
  a specialised drip torch, using a gelled fuel, slung and activated from a helicopter.
- **hemicellulose**
  this wood cell material is made from chains of sugar molecules and accounts for about 20% of the dry weight of wood.

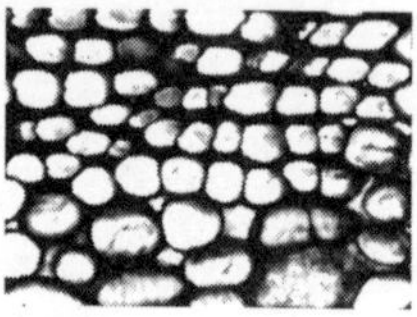

- **herbaceous**
  vegetation or parts of plants with little or no woody tissue.

- **herbicide**
  chemical substances or living organisms (called bioherbicides) used to kill or control vegetation such as

brush, weeds, and competing or undesirable trees.

- **herbicide resistant**
capable of surviving herbicides that previously would have killed the plants along with the targeted weeds.

- **herbivore**
an organism that eats plants.

- **heritage areas**
sites of historical, architectural, archaeological, paleontological or scenic significance to the province.

- **heritage trail**
a trail having cultural significance by reason of established aboriginal use or use by early immigrants.

- **herringbone felling**
timber felled in a herringbone pattern towards the center of the yarding road.

- **heteroecious**
requiring two taxonomically different hosts to be able to complete the entire life cycle, as in the rust fungi.

- **heterokont**
a term used variously to refer to some or all of those algae with chloroplasts having chlorophylls a and c (i.e. stramenopiles, cryptomonads, haptophytes and dinoflagellates). The group is paraphyletic as it fails to include some taxa derived from heterokonts.

- **heterotrophic**
a mode of nutrition in which the consumer relies upon molecules created by other organisms for energy and nutrients. Either osmotrophic (absorbing soluble organic matter) or phagotrophic (ingesting particles of food).

- **high climber**
a logger who climbs trees to top them or rig them with blocks for cable yarding. Also known simply as a 'climber' or 'tree climber'. Although few, if any, landing spar trees are rigged any more (portable towers having replaced them), there is still a need for climbing trees to rig them for tail spars or to set lines for directional tree pulling during felling.

- **high flotation**
equipment that has a ground pressure of less than five or six pounds per

Square Inch (PSI). Less than four PSI may be needed to operate on swamps without difficulty.

- **high grade**

good quality timber. To cut the cream of the crop; take only the best trees out of the stand.

- **high grading**

removing all mature, good quality trees from a stand, while leaving inferior trees and less desirable species uncut. Natural regeneration from the retained trees will likely produce a poor-quality stand. High grading should be distinguished from uneven-aged management wherein some (but not all) trees in all diameter classes are removed in order to create a high-quality stand.

- **high hazard (forest health)**

physical characteristics (including tree species, composition, age, and size) and biogeoclimatic factors that make a forest highly susceptible to attack by damaging agents.

- **high sensitivity areas**

areas having special concerns, issues, or the potential for negative impacts on resource values, including any soils with high hazard or very high hazard for compaction, erosion, mass wasting, or displacement.

- **highgrading**

1. an undesirable timber harvest practice that 'takes the best and leaves the rest.'
2. the removal of only the best trees from a stand, often resulting in a residual stand of poor quality trees.
3. a harvesting technique that removes only the biggest and most valuable trees from a stand and provides high returns at the expense of future growth potential. Poor quality, shade-loving trees tend to dominate in these continually highgraded sites.
4. a type of timber harvesting in which larger trees of commercially valuable spe-

cies are removed with little regard for the quality, quantity, or distribution of trees and regeneration left on the site; often results when a diameter limit harvest is imposed.

- **highlead**
  a cable yarding system in which lead blocks are hung on a spar or tower to provide lift to the front end of the logs.

- **high-lead logging**
  wire rope system that involves yarding in logs or trees by means of a rope passing through a block at the top of the head spar.

- **highlead system**
  logging system that uses cables rigged to a spar high above the ground so that one end of the logs can be lifted during yarding.

- **high-performance linerboard**
  papers with the same strength as regular paper grades but produced at a lower weight. This process gives more compression strength per pound of fiber than conventional liner, thus reducing total fiber use.

- **hip chain**
  a device used to measure distance by means of an anchored filament wrapped around a wheel that revolves as you walk (handy for measuring distances on your own).

- **historical variation**
  the range of the spatial, structural, compositional and temporal characteristics of ecosystem elements during a period specified to represent 'natural' conditions.

- **hoe-chucking**
  a logging system that uses an excavator or hoe to yard logs to the roadside and/or landing.

- **hoedag**
  hoe like tool with an elongated blade used for planting trees.

▪ **hog**
machine used to grind wood into chips for use as fuel or for other purposes; the wood used is usually waste wood unfit for lumber or other uses. Coarse wood chips to be burned as a fuel.

▪ **hog fuel**
coarse wood chips and bark produced to be burned as fuel. Also known as 'biomass chips'.

▪ **hog wood**
pulpwood logs to be chipped in a hog.

▪ **hogged fuel**
fuel made by grinding waste wood in a hog; a mix of wood residues such as sawdust, planer shavings and sometimes coarsely broken-down bark and solid wood chunks produced in the manufacture of wood products and normally used as fuel.

▪ **holding wood**
when felling timber, this is the part of a tree left uncut until the end in order to hold the fall of the tree in the desired direction.

▪ **holopelagic**
organisms that remain pelagic throughout their entire life cycle.

▪ **homology**
a characteristic shared by different groups of organisms that is derived from the same structure or trait in their common ancestor. Ancestral characteristics may be greatly modified in descendent groups, so that homologous traits may not be similar in appearance or function. Clues of their common ancestry may often be found in structural details, ontogeny, or position relative to other features.

▪ **hooktender**
the working foreman in charge of a yarding crew.

▪ **hoot owl**
a logging operation limited to the early morning hours when fire danger later in the day will require a cessation of operations.

▪ **horizontal gene transfer**
a natural process by which some of the genes of an organism are transferred and incorporated into the genome of an organism be-

longing to another species. This contrast with vertical gene transfer, in which the genes of an organism are passed on to its offspring.

- **horizontal structure**
  the spatial arrangement of plant communities; a complex horizontal structure is characterised by diverse plant communities within a given geographic unit.

- **horse logging**
  logging with large draft horses. Very specialised form of logging where the main advantage is low stand impact in selection logging.

- **host**
  a living organism serving as a food source for a parasite.

- **host specific**
  capable of surviving only in a relationship with a particular host species.

- **hot deck**
  a pile of logs from which logs are hauled as soon as they are yarded. AKA 'hot loading' or 'to hot log'.

- **hot pond**
  a place to soak logs in water to remove dirt and to loosen the barkdimension

an integral component of ecosystem management that recognises people are part of ecosystems, that people's pursuits of past, present and future desires, needs and values (including perceptions, beliefs, attitudes and behaviours) have and will continue to influence ecosystems and that ecosystem management must include consideration of the physical, emotional, mental, spiritual, social, cultural and economic well-being of people and communities.

- **human impact or influence**
  a disturbance or change in ecosystem composition, structure or function caused by humans.

- **humus**
  1. a general term for the more or less decomposed plant and animal residues in the lower organic soil layer. 2. the dark organic part of soil formed from decaying plant and animal matter, often called topsoil.

- **hybridisation**
  the process of sexual reproduction using genetically distinct parents.

■ **hydric soils**
saturated soils that typically have little oxygen content.

■ **hydrocarbon**
an organic compound containing only carbon and hydrogen, often occurring in petroleum, natural gas and coal.

■ **hydrologic unit code**
an inventory geographical locator for water uses and activities affecting watershed response and fisheries.

■ **hydrology**
the science that describes and analyses the occurrence of water in nature, and its circulation near the surface of the earth.

■ **hydrolysis**
1. the breaking apart of a molecule by addition of water.
2. water-tolerant plants characteristic of wetlands.

■ **hydrophytic**
plants that grow in and are adapted to an aquatic or very wet environment.

■ **hydroseeding**
the application of seed in a water slurry that contains fertiliser, a soil binder and/or mulch.

■ **hypha**
microscopic, tubular filaments that make up the body of a multicellular fungus.

■ **hypochondria**
a term used to refer to premitochondriate eukaryotes.

■ **hypsometer**
a simple instrument, often a stick or other straight edge, used to measure the heights of trees on the basis of similar angles.

■ **hysterothecium**
a sexual fruiting structure of the Ascomycete fungi, usually football-shaped or elongate in appearance and occurring on infected needles.

■ **ice age**
the Pleistocene period when a large part of the Northern Hemisphere was intermittently covered with glaciers.

■ **idiosome**
a structure produced by the organism, as opposed to a xenosome or foreign body. Used to refer to the ele-

ments which make up or adhere to the test of some amoebae.

- **immature**
  trees or stands that have grown past the regeneration stage, but are not yet mature.

- **immature timber**
  stands of timber where the age of the leading species in a stand is less than the specified cutting age. Cutting ages are established to meet forest management objectives. Usually stands with lodgepole pine and whitebark pine or a deciduous species as the leading species are considered as immature timber when the stand age is less than 81 years. Otherwise, all stands having conifers other than lodgepole pine and whitebark pine as the leading species are immature when the stand age is less than 121 years.

- **immotile**
  not moving, sometimes used to refer to cells which are fixed to the substrate.

- **impact assessment**
  a study of the potential future effects of resource development on other resources and on social, economic and/or environmental conditions.

- **improved plow steel**
  the standard grade of wire rope with a breaking strength 15% less than EIPS.

- **improvement cut**
  an intermediate cut made to improve the form, quality, heath or wildlife potential of the remaining stand.

- **improvement cutting**
  the removal of trees of undesirable species, form or condition from the main canopy of the stand to improve the health, composition and value of the stand.

- **in vitro**
  the use of an artificial environment, such as a test tube.

- **in vivo**
  the use of a natural environment, such as within a living organism.

- **incentive**
  a reward for improving forest management. Incentives include reimbursement of some expenses but can also take the form of an abate-

ment of property or income tax.

- **incentive**

  a reward for improving forest management. Incentives include reimbursement of some expenses but can also take the form of an abatement of property or income tax.

- **incising**

  perforating the wood surface with small slits to improve the penetration and effectiveness of the preservatives. Not necessary with treated Southern Yellow Pine, Ponderosa Pine and Red Pine.

- **incising**

  perforating the wood surface with small slits to improve the penetration and effectiveness of the preservatives. Not necessary with treated Southern Yellow Pine, Ponderosa Pine and Red Pine.

- **increment**

  the increase in diameter, basal area, height, volume, quality or value of individual trees or stands during a given period.

- **increment borer**

  a tool used to extract a core of wood from a living tree for the purpose of studying the annual growth rings of the tree.

- **increment core**

  that part of the cross section of a tree extracted by an increment borer. Used to determine tree age and growth pattern.

- **indicator species**

  species with such specialised ecological needs that they can be used for assessing the quality, condition, or extent of an ecosystem on the basis of their presence and density, or the accumulation and effect of materials in their tissues.

- **indigenous**

  pertaining to plants or animals that are native to a particular region or country.

- **industrial operation**

  operations such as land clearing, timber harvesting, timber processing, mechanical site preparation and other silvicultural treatments, mining and road construction.

- **inhaul**
  the portion of a cable yarding cycle where a turn of logs is brought to the landing.

- **initial attack**
  the action taken to halt the spread or potential spread of a fire by the first fire fighting force to arrive at the fire.

- **initial mature inventory**
  that portion of the existing total mature forest inventory which is available for harvest. This portion reflects all management constraints that are necessary to protect the environment and other forest uses and varies with the constraints identified for each option.

- **inner gorge**
  a stream reach or portion of stream that is bounded by steep hillslopes (> 40% sideslope) and terminates upslope into more gentle topography.

- **inoperable lands**
  lands that are unsuited for timber production now and in the foreseeable future by virtue of their elevation; topography; inaccessible location; low value of timber; small size of timber stands; steep or unstable soils that cannot be harvested without serious and irreversible damage to the soil or water resources; or designation as parks, wilderness areas or other uses incompatible with timber production.

- **inorganic**
  composed of matter that is not animal or vegetable; not having the organised structure of living things.

- **insecticide**
  any substance used to kill insects.

- **insloping**
  shaping the road surface to direct water onto the cut side of the road. Integrated Resource Management (IRM): the identification and consideration of all resource values, including social, economic, and environmental needs, in land use

and development decision making. It focusses on resource use and land use and management, and is based on a good knowledge of ecological systems, the capability of the land and the mixture of possible benefits.

- **instar**
  the insect itself during the time between molts in the larva or nymph, numbered to designate the various periods; i.e., the first instar is the insect between the egg and first molt.
- **instream cover**
  the amount of vegetation and organic debris within a stream channel capable of providing protection for fish.
- **instream woody debris**
  a measure of the amount, size and arrangement of woody material in a stream.
- **integrated forest management practices**
  the appropriate mix of activities aimed to achieve forest conditions of desired value(s).
- **Integrated Pest Management (IPM)**
  an ecological approach to pest management in which all available necessary techniques are consolidated into a unified program so that pest populations can be managed in such a manner that economic damage is avoided and adverse side effects are minimised.
- **integrated resource use**
  a decision making process whereby all resources are identified, assessed and compared before land use or resource management decisions are made. The decisions themselves, whether to approve a plan or carry out an action on the ground, may be either multiple or single use in a given area. The application of integrated resource management results in a regional mosaic of land uses and resource priorities which reflect the optimal allocation and scheduling of resource uses.
- **integrated manufacturer**
  an integrated forest products company grows its own trees, makes products from them, makes other products from fiber leftovers from the

initial manufacturing, then converts and adds value to all these products.

- **intensive forest management**
utilisation of a wide variety of silvicultural practices, such as planting, thinning, fertilisation, harvesting and genetic improvement, to increase the capability of the forest to produce fibre.

- **interior conditions**
at a point where edge effects no longer influence environmental conditions within a patch, interior conditions are achieved. For coastal B.C. forests, the edge effect is generally felt for a distance equivalent to 2 to 4 times average tree height into the stand. The effects usually involve light intensity, temperature, wind, relative humidity and snow accumulation and melt. See **edge effect**.

- **interlock yarder**
a yarder which incorporates a means of coupling the main and haulback drums to recirculate the horsepower required to maintain running line tensions.

- **intermediate**
intermediate trees have crowns below, but still extending into, the general level of the canopy and receive a little direct light from above but none from the sides.

- **intermediate trees**
trees or shrubs receiving little direct light from above, and none from the sides; usually with small crowns considerably crowded on the sides that are generally either below or extending into the canopy formed by codominant trees or shrubs.

- **intermittent stream**
1. stream that flows primarily during the wet season but has little or no flow during other seasons.
2. a stream or portion of a stream that flows only in direct response to precipitation. It is dry for a large part of the year.

- **international 1/4 inch scale**
log scale modification of an earlier rule using a 1/8-inch kerf, based on an analysis of the loss of wood fibre incurred in the conversion of saw logs to lumber. One of

the few rules incorporating a basis for dealing with log taper.

- **interplant**
to plant between or among other plants; to plant more than one species on the same site

- **interpretive forest site**
a designated forest site and ancillary facilities developed by the Ministry of Forests to interpret, demonstrate, or facilitate the discussion of the natural environment, forest practices and integrated resource management.

- **intertree distance**
the distance between tree boles, usually used in the context of thinning. Recommended guidelines for intertree distances are established for different thinning programs depending on site variables, the species and age of trees and management objectives.

- **interveinal**
between veins.

- **intolerant**
tree relatively incapable of developing and growing normally in the shade of and in competition with, other trees.

- **intraveinal**
associated along or within veins.

- **invaders**
plants which invade a site after fire.

- **inventory**
1. an itemised list or catalogue of goods, property, etc.
2. a survey of a forest area to determine such data as area, condition, timber, volume and species for specific purposes such as planning, purchase, evaluation, management or harvesting.

- **invertebrates**
an animal lacking a backbone. In streams, commonly encountered invertebrates include crayfish, mussels, worms and insects.

- **invertebrates**
an animal lacking a backbone. In streams, commonly encountered invertebrates include crayfish, mussels, worms and insects.

- **jackpot**
a pile of haphazardly felled trees.

- **jagger**
  a broken strand projecting out of a worn wire rope.

- **jammer**
  a light weight, ground lead yarder using tongs and usually mounted on a truck with a spar and boom. Usually relatively inexpensive and equipped with a single drum. It may be used for both yarding and loading.

- **jill-poke**
  a supported straight log, pole, sapling, or bar that strikes an object at an angle, either penetrating the object or pushing it a side. Cat skinners and skidder operators must be especially careful of jill-pokes while skidding.

- **juku**
  private Japanese tutorial schools that supplement school learning, offer remedial and enrichment experiences and prepare students for university examinations.

- **jungle**
  land covered with dense growth of trees, tall vegetation, vines, etc., typically in tropical regions, and inhabited by predatory animals.

- **juvenile wood**
  inner core of xylem surrounding the pith. The cells are smaller and less structurally developed than those of

the outer xylem. The time during which juvenile wood is formed is termed the juvenile period. This period varies among individuals, with species, and with environmental conditions.

- **KDAT**
  wood that is Kiln-Dried after Treatment. Pressure-treating with water-borne preservatives adds moisture to the wood, which can be removed by kiln-drying.

- **kerf**
  the width of a saw's cut.

- **key area**
  a relatively small area selected because of its location, use, or grazing value as a monitoring point for grazing use. It is assumed that key areas, if properly selected, will reflect the overall acceptability of current grazing management.

- **key species**
  forage species that must, because of their high degree of use, be considered in the management program.

- **keystone species**
  a species that plays an important ecological role in determining the overall structure and dynamic relationships within a biotic community. A keystone species presence is essential to the integrity and stability of a particular ecosystem.

- **keystone species**
  species that play roles affecting many other organisms in the ecosystem.

- **kg and pile**
  a site preparation method in which stumps are pushed up, sheared off, or split apart by a specially designed blade mounted on a bulldozer. Debris is then piled or placed in long rows (windrows) so that an area can be bedded or flat planted.

- **kg blade**
  1. a bulldozer-mounted blade used in forestry and land-clearing operations. A single spike splits and shears stumps at their base.
  2. a type of specialised bulldozer blade used in forestry site-prep applications to cut through leftover stumps and pile large debris to help clear land for tree regeneration.

- **kickback**
  1. when a chainsaw's bar rebounds upward off the log being sawed, often in a uncontrolled manner and out of the faller's control.
  2. when a tree being felled slips backward off the stump toward the faller.

- **kiln**
  a furnace or oven for drying, burning, or baking something, as bricks, grain or pottery.

- **kip**

a unit of weight or force equal to 1,000 pounds; 'thousand (kilo) in pounds.'

- **knuckleboom**

hydraulically operated loading boom whose mechanical action imitates the human arm.

- **kraft**

1. a very strong type of wrapping paper made from a special type of chemically treated wood pulp. The Kraft (Sulfate) pulping process produces the best quality softwood pulp. Generally brown natural kraft but can be bleached white. Kraft is the German word for strength.

2. a German word that means strong, it is the strongest and most widely used grade of pulp made today.

- **kraft paper**

comparatively coarse paper particularly noted for its strength; unbleached grades are used primarily as a wrapping or packaging material. Paper made primarily from wood pulp produced by the sulphate pulping process.

- **k-strategist**

species that produce a few, often large offspring, but invest a great deal of time and energy to ensure that most of the offspring will reach reproductive age.

- **ladder fuels**

fuels that provide vertical continuity between the surface fuels and crown fuels in a forest stand, thus contributing to the ease of torching and crowning.

- **lake**

a naturally occurring static body of water greater than 2 m in depth and greater than 1 ha in size or a licensed reservoir.

- **lamina**

the wide part of the leaf; also called the leaf blade.

- **laminated linerboard**

two or more plies of linerboard adhered to one another for increased structural stability.

- **land**

  to place logs into a landing during yarding or skidding operations.

- **Land And Resource Management Plan (LRMP)**

  a strategic, multi-agency, integrated resource plan at the sub regional level. It is based on the principles of enhanced public involvement, consideration of all resource values, consensus-based decision making and resource sustainability.

- **land cover category**

  that which overlays or currently covers the ground, especially vegetation, permanent snow and ice fields, water bodies, or structures. Barren land is also considered a 'land cover' although technically it is lack of cover. The term land cover can be thought of as applying to the setting in which action (one or more different land uses) takes place. Classes include Built-up land, Cultivated Cropland, Grassland, Shrubland, Treeland, Water areas and Barren land.

- **land ethic**

  the principles and values guiding our use and treatment of the land. Forest stewardship is a land ethic.

- **land location**

  area identifiers composed of meridian, township, range, section and subdivision.

  Meridian: Two lines, one extending north and south along the astronomical meridian and the other east and west along a true parallel of latitude, intersecting at the initial point, along which township, section and quarter-section corners are established. The meridian and Base line are the lines from which the survey is initiated for the township boundaries along the parallels.

  Township: The unit of survey of the public lands; normally a quadrangle approximately 6 miles on a side with boundaries conforming to meridians and parallels within established limits, containing thirty-six sections, some of which are designed to take up the convergence of the meridianal township boundary lines and accumulated measurement errors.

Range: Any series of contiguous townships situated north and south of each other; also sections similarly situated within a township.

Section: The unit of subdivision of a township; normally a quadrangle 1 mile square, with boundaries conforming to meridians and parallel within established limits, and containing 640 acres.

Subdivision: The sixteen standard forty-acre aliquot parts of a section.

- **land use class**

the predominant purpose for which an area is employed. Classes include Agricultural Land, Forest land, Rangeland, Wetland, Urban/suburban and Utility/Transportation Corridors (Roads, Railroads, Utility Corridors).

- **land/aquatic type association**

code numbers given to a mapped unit of land in which land forms, soils, vegetation and water have the dominating influence.

- **landforms**

any physical feature of the earth's surface having a characteristic, recognisable shape and produced by natural causes. Landforms include:

Alluvial Fan: A body of unconsolidated plastic material and debris flow, conical in shape, forming at the point where a stream emerges from a narrow valley onto a broader, less sloping valley floor.

Backslope: The component of the hill slope that forms the steepest inclined surface and is frequently the principal element. The surface is dominantly steep and linear in profile and erosional in origin.

Badlands: Intricately stream-dissected topography characterised by a very fine drainage network with high drainage densities and short, steep slopes. They have little or no vegetative cover overlying consolidated or poorly cemented clays or silts.

Bajada: A broad, gently inclined slope formed by the lateral coalescence of a series of alluvial fans, and having a broadly undulating transverse profile.

Basin: A depressed area with no surface outlet.

Butte: An isolated, usually flat-topped upland mass characterised by summit widths that are less than heights of the bounding erosional escarpment. It is produced by differential erosion of nearly horizontal, interbedded weak and resistant rocks. See also **mesa**.

Caldera: A large, basin-shaped volcanic depression, more or less circular in form, and having a diameter many times greater than the included volcanic vents. See also **crater**.

Canyon: A long, deep, narrow, very steep-sided valley with high and precipitous walls in an area of high local relief.

Cirque: Semicircular, concave, bowl-like area with steep face primarily resulting from erosive activity at the head of a mountain glacier.

Cliff: A high, very steep to perpendicular or overhanging face of rock or earth.

Cone: A conical hill of lava or cinders that is built up around a volcanic vent.

Crater: A basin-like rimmed structure, usually at the summit of a cone; its floor is approximately the diameter of the vent. See also **caldera**.

Divide: The line of separation marking the boundary between two adjacent drainage systems.

Draw: A small stream channel, generally more open and with a broader floor than a gulch. Also locally called 'arroyo' or 'wash'.

Dune: A mound, ridge, or hill of loose, wind-blown granular material (generally sand), either bare or covered with vegetation.

End Moraine: The moraine produced at the front of an actively flowing glacier at any given time.

Escarpment: A relatively continuous and steep slope or cliff breaking the general continuity of more gently sloping land surfaces and produced by erosion or faulting. When applied to cliffs formed by faulting, commonly abbreviated to 'scarp.'

Flat: A level or nearly level area of land marked by little or no relief.

Flood plain: The nearly level alluvial plain that borders a stream and is subject to inundation under flood stage conditions.

Foot slope: The component of the hill slope that forms the inner, gently inclined surface at the base. The surface is dominantly concave in profile and is transitional between erosion and deposition.

Gorge: A narrow, deep valley with nearly vertical rocky walls; used especially to identify a restricted, steep-walled part of a canyon.

Ground Moraine: An extensive, fairly even and undulating layer of rock debris which has been primarily deposited from underneath the glacier.

Gulch: A small stream channel, narrow and steepsided in cross section. Also called 'ravine.'

Hill: A natural elevation of the land surface, rising as much as 1000 ft (see **mountain**) above the surrounding lowlands, usually of restricted summit area and having a well-defined outline. Hills fringing a mountain range are called 'foothills.'

Hill slope: The steeper part of a hill between its summit and the drainage line or valley floor. Components of the hill slope may be further classified as shoulder, backslope, footslope, and toe slope.

Hogback: A sharp-crested, symmetric ridge formed by the differential erosion of highly tilted and resistant rock layers.

Karst: A type of topography characterised by closed depressions, sinkholes, caves, and underground drainage. It is formed by the underground solution of limestone, dolomite, and other soluble rocks and the associated processes of subsidence and collapse.

Knob: A small, rounded hill, commonly isolated or rising above adjacent landforms. Also called 'knoll'.

Lateral Moraine: A ridge-like moraine carried on and deposited at the side margin of a valley glacier.

Mesa: A broad, nearly flat-topped, and usually isolated

upland mass characterised by summit widths that are greater than the heights of bounding escarpments. As summit area decreases relative to height, mesas are transitional to buttes.

Moraine: An accumulation of rock material, with an initial topographic expression of its own, built chiefly by the direct action of glacial ice or by running water emanating from the glacier. Moraines may be classified as end moraine, ground moraine, lateral moraine, recessional moraine, or terminal moraine depending on their relationship to the movement of the ice mass.

Mountain: A natural elevation of the land surface, rising more than 1000 ft (see hill) above the surrounding lowlands, usually of restricted summit area (see plateau), and generally having steep sides.

Peak: Sharp or rugged upward extension of a ridge chain, usually at the junction of two or more ridges; the prominent highest point of a summit area.

Piedmont: An area or feature at the base of a mountain or mountain range.

Plain: An extensive lowland area that ranges from level to gently sloping or undulating.

Plateau: An extensive upland mass with a relatively flat summit area that is considerably elevated above adjacent lowlands, and is separated from them on one or more sides by escarpments. A comparatively large part of its total surface is at or near the summit level. See mesa and mountain.

Playa: The usually dry and nearly level lake plain that occupies the lowest parts of a closed basin.

Recessional Moraine: An end moraine, built during a temporary but significant halt in the final retreat of a glacier.

Ridge: A long, narrow elevation of the land surface, usually sharp crested with steep sides and forming an extended upland between valleys.

Saddle: A low point on a ridge or crestline, generally a divide between the heads of streams flowing in opposite directions.

Shoulder: The component of the hill slope that forms the uppermost inclined surface. The surface is dominantly convex in profile and erosional in origin.

Stream Terrace: One of a series of platforms in a stream valley, flanking and generally parallel to the stream channel, originally formed near the level of the stream, and representing the dissected remnants of an abandoned flood plain produced during a former stage of erosion or deposition.

Structural Bench: A platform-type, nearly level to gently inclined erosional surface developed on resistant strata in areas where valleys are cut in alternating strong and weak layers with an essentially horizontal attitude.

Summit: A general term for the top, or highest level, of the relatively undissected upland between two adjacent valleys.

Terminal Moraine: An end moraine that marks the farthest advance of a glacier and usually has the form of a massive arcuate ridge, or complex of ridges.

Toe Slope: The component of the hill slope that forms the outermost, gently inclined surface at the base. The surface is dominantly linear in profile and depositional in origin.

Valley: An elongate, relatively large, externally drained, gently sloping depression of the Earth's surface commonly situated between two mountains or ranges of hills or mountains. It is usually developed by stream erosion.

Valley Floor: The gently sloping to nearly level bottom surface of a valley.

- **landing**

1. an area modified by equipment that is designed for accumulating logs before they are transported.
2. any place on or adjacent to the logging site where logs are assembled for further transport.
3. area on a harvested site where wood is transferred from the skidders or forwarders to trucks for road transport. In some opera-

tions the trees are processed or stored on the landing before transport.

- **landing man**
worker who bucks, limbs, and trims log ends, unhooks chokers and assists in hooking up trailers to logging trucks. Also see 'Chaser'.

- **landing pile or cull pile**
an area of piled slash, logging residue, and stumps, created as a result of harvesting operations and the construction of roads and landings.

- **landscape**
a heterogenous land area with interacting ecosystems that are repeated in similar form throughout.

- **landscape diversity**
the size, shape and connectivity of different ecosystems across a large area.

- **landscape ecology**
the study of the distribution patterns of communities and ecosystems, the ecological processes that affect those patterns and changes in pattern and process over time.

- **landscape features**
the land and water form vegetation, and structures that compose the characteristic landscape.

- **landscape level**
a watershed, or series of interacting watersheds or other natural biophysical (ecological) units, within the larger Land and Resource Management Planning areas. This term is used for conservation planning and is not associated with visual landscape management and views cape management.

- **landscape sensitivity**
a component of the landscape inventory that estimates the sensitivity of the landscape based on the visual prominence of importance of features; conditions that affect visual perception; and social factors that contribute to viewer perceptions.

- **landslide**
a movement of earth down a steep slope.

- **land-use planning**
the process by which decisions are made on future land uses over extended time periods, that are deemed to best serve the general welfare.

- **lang lay**
a type of wire rope more suitable for running through multiple sheaves. The wires making up the strands run the same direction as the strands themselves, giving greater flexibility.

- **Large Organic Debris (LOD)**
entire trees or large pieces of trees that provide channel stability or create fish habitat diversity in a stream channel.

- **large woody debris**
1. a large tree part, conventionally a piece greater than 10 cm in diameter and 1 m in length.
2. pieces of wood larger than 10 feet long and 6 inches in diameter, in a stream channel.

- **larva**
the immature instars, between the egg and pupal stages, in an insect having a complete metamorphosis (egg, larva, pupa, adult). Larvae feed and grow but cannot fly, nor can they reproduce.

- **late seral state forest**
stage in forest development that includes mature and old-growth forest.

- **lateral yarding**
any movement of logs toward the centerline of a yarding road.

- **lateral yarding distance**
the maximum distance perpendicular to each side of a skyline road within which logs can be attached for yarding.

- **lay**
1. the position of a felled tree on the ground.
2. the direction that strands of a wire rope are spirally wound around the core.

- **layout**
1. a logging plan including units and roads.
2. the position of the running lines in a cable yarding system.

- **leaching**
downward movement of a pesticide or other soluble

material through the soil as a result of water movement.

- **lead**

1. the direction of the operating or main line(s).
2. a block or series of blocks or rollers attached to a stationary object to guide the cable by which logs are dragged.
3. the established direction in which all the trees on a given strip or setting are to be felled.

- **lead, square**

a yarding layout where the main line make a 90° angle with the axis of the yarder frame. Usually associated with spar trees where the yarder sled or frame is separate from the spar. Term is not useful with modern portable, standing spar towers.

- **lead, straight**

a yarding layout where the mainline is inline with the axis of the yarder and spar frame. Usually associated with spar trees where the yarder sled or frame is separate from the spar. Term is not useful with modern portable, standing spar towers.

- **leader**

the length of tree stem from the top of the tree down to the first set of branches, representing one year of growth and reflecting the tree's vigour and the site's growing potential.

- **leadership**

the quality which enables a person to articulate the current reality and a vision for the future. The ability to mobilise a community of interest behind the vision. People will follow leaders where they wouldn't go on their own.

- **leaflet**

one of the subdivisions of a compound leaf.

- **lean**

the direction in which a tree or trees are leaning.

- **leave strip**

strip of uncut timber left between cutting units or adjacent to another resource such as a stream. Also known as a buffer strip, green strip, or streamside management zone.

- **leave trees**

all trees, regardless of species, age, or size, remaining on a harvested area as a result of a predetermined silviculture prescription to address a possible range of silviculture or resource needs.

- **legacy**
  how you will be remembered, and how your organisation will be remembered. Writing your own legacy is an excellent way to come up with a vision statement. Once you know how you want to be remembered, you have a vision and can go to work on making it a reality. Obviously, you don't want to wait too long to do this.

- **legumes**
  plants that produce organic nitrogen from nitrogen gas in the air. These plants, which typically form seeds in pods, include soybeans, peas, alfalfa, lespedeza and locust.

- **length of ground**
  used to describe whether the ground in a unit is badly cut up and rough or smooth and uniform.

- **lesion**
  a wound; a well-marked, but limited, diseased area; a break or rupture through a tissue, especially a surface tissue.

- **life cycle**
  the phases, changes, or stages through which an organism passes during its lifetime.

- **lifelong learning**
  the ongoing acquisition of knowledge or skills.

- **lightning detection system**
  a network of electronic field sensors linked to a central computer to detect, triangulate, plot the location of and record cloud-to-ground lightning flashes in real time over a predetermined area.

- **lignin**
  1. a polymer that binds cellulose together to make wood. The stiffening material inside wood cell walls that gives strength and rigidity. Allows trees to grow tall and out-compete other plants for sunlight. Accounts for about 30% of the dry weight of wood. Lignin is removed in the pulping and bleaching processes to brighten the paper and prevent yellowing.

2. an organic substance which acts as a binder for the cellulose fibres in wood and certain plants and adds strength and stiffness to the cell wall.

- **limb**
  to remove the limbs from a felled tree. To cut branches from trees or logs.

- **limbwood**
  part of the tree above the stump that does not meet the requirement for saw logs or upper stem portions. Includes all live, sound branches to a 4-inch outside bark diameter minimum.

- **limiting factor**
  a factor present in an environment in such short supply that it limits growth or some other life process.

- **limnology**
  the study of the biological and chemical properties of fresh water.

- **line pulling**
  felling trees against the lean by securing a line and pulling it over with a cat or line winch. This procedure is commonly required in stream protection areas. A form of 'Directional felling'.

- **line swivel**
  a swivel in the butt rigging or line.

- **lineage**
  a group of organisms linked to one another through a continuous line of descent, i.e., through parent-offspring connections. Biologists talk of lineages with respect to populations of organisms, cells or genes.

- **linear developments**
  straight line industrial development that is typical of power lines, highways, gas lines and seismic activities.

- **linerboard**
  the inner and outer layers of paper that form the wall of a corrugated container.

- **lithologic unit**
  a system of rock classification based on manner of origin, composition, and texture (or grain size). No one system is universally recognised; the following nomenclature is offered as a guide to common terms that would be significant in most Forest Service work.
  Alluvium: Deposited by running water.

Amphibolite: A dark-coloured, medium-grained rock containing amphibole and plagioclase.

Andesite: Extrusive equivalent of diorite

Anhydrite: Anhydrous (does not contain water) calcium sulfate, similar to gypsum but harder and slightly less soluble.

Aphanitic Texture: Individual components can not be identified with the unaided eye.

Arkose: Feldspar-rich (usually greater than 25%) sandstone that is not well-sorted.

Asphalt: A dark brown or black, low melting point, bitumen (a natural inflammable substance) comprised almost entirely of carbon and hydrogen.

Basalt: Extrusive equivalent of gabbro.

Breccia: Consolidated gravel composed of broken, angular particles.

Carbonaceous: Rich in carbon or organic matter.

Carbonate: A general rock type formed from organic and inorganic precipitation of calcium and magnesium carbonates.

Chert: A dense, very hard rock composed of microcrystalline silica (varieties also called flint, jasper, agate)

Coal, Anthracite: A very hard, black coal, actually classed as metamorphic.

Coal, Bituminous: A harder, more compacted, black coal.

Colluvium: Deposited by rainwash or slow downslope creep.

Conglomerate: Consolidated gravel composed of rounded particles.

Dacite: Extrusive equivalent of granodiorite.

Diatomite: A light-coloured, soft rock composed of the siliceous skeletons of diatoms (water-dwelling organisms).

Diorite: Plagioclase (Sodium-rich) feldspar, quartz less than 10%, magic minerals about 25%

Dolomite: Contains more than 90% of the mineral dolomite (calcium-magnesium carbonate).

Eolian: Deposited by wind.

Evaporite: A general group of rocks produced by the extensive evaporation of a saline solution.

Foliate: A general term for planar or layered structure

Gabbro: Plagioclase (Calcium-rich) feldspar about 50%, magic minerals about 50%.

Glacial: Deposited by action of glaciers or ice sheets.

Glassy Texture: Amorphous rock without distinct crystallisation.

Gneiss: A rock with alternating bands of granular and flaky (or elongate) minerals. Generally less than 1/2 the minerals show a preferred parallel orientation.

Granite: Orthoclase feldspar greater than 2/3 total feldspar, quartz greater than 10%.

Granodiorite: Plagioclase (Sodium/Calcium) feldspar greater than 2/3 total feldspar, quartz greater than 10%.

Gravel: Particle size greater than 2mm.

Graywacke: A dark-grey, coarse-grained, poorly-sorted sandstone.

Gypsum: A soft mineral consisting of hydrous calcium sulfate.

Halite: Native salt.

Hematite: A common iron oxide mineral, occurring in several forms, with a distinctive brick-red colour when powdered.

Hornfels: A fine-grained rock with a mosaic of equidimensional grains.

Igneous: A rock that has solidified from molten or partly molten material (magma).

Lacustrine: Deposited on the bottom of a lake.

Latite: Extrusive equivalent of monzonite.

Lignite: A soft, brown coal formed by the further compression of peat.

Limestone: Contains more than 95% calcite (calcium carbonate).

Limonite: A general group of hydrous (contains water) ferric oxides, commonly having a dark brown or yellow-brown colour.

Man-made: Landfills and earthworks.

Marble: Recrystallised calcite and/or dolomite, usually with a sugary texture.

Marine: Deposited on the bottom of a sea.

Massive: A general term denoting lack of foliation.

Metamorphic: Rock that has been derived from pre-existing rocks, essentially in the solid state, in response to changes in temperature, pressure, shearing stress, and chemical environment.

Metaquartzite: Recrystallised sandstone or chert, consisting mainly of quartz.

Migmatite: A composite rock containing both igneous and metamorphic materials.

Monzonite: Orthoclase/Plagioclase ratio about equal, quartz less than 10%.

Mylonite: A compact, chertlike rock without cleavage, but with a banded appearance produced by extreme shearing and pulverising during metamorphism.

Obsidian: Black to dark-coloured volcanic glass, similar in composition to rhyolite.

Organic: Accumulations of organic matter.

Peat: A soft, brown material containing the partially decomposed remains of plants.

Pegmatitic Texture: Exceptionally coarsely crystalline, usually with a composition similar to granite. Commonly referred to as 'pegmatite'.

Peridotite: Magic minerals greater than 90%.

Phaneritic Texture: Individual components can be identified with the unaided eye.

Phosphorite: Rock containing quantities of precipitated or reworked phosphate minerals.

Phyllite: Contains platy minerals too small to be clearly identifiable, distinguished by a glossy sheen.

Porphyritic Texture: Larger crystals set in a finer-grained groundmass. Takes name from dominant rock type.

Pumice: Light-coloured, vesicular (filled with small cavities formed by entrapment of gases) glassy rock, similar in composition to rhyolite.

Rhyolite: Extrusive equivalent of granite.

Sand: Particle size between 1/16 and 2mm.

Sandstone: Cemented sand. When used without a qualifier, generally contains about 90% quartz.

Schist: Can be readily split into thin flakes or slabs due to more than 1/2 the minerals showing a well-developed parallelism.

Sedimentary: Rock that has formed from the accumulation of materials on the Earth's surface.

Serpentine: A rock with a greasy or silky lustre and a tough, conchoidal fracture, having a common greenish colour and often veined or spotted.

Shale: Finely laminated, clayey rock with about 1/2 silt.

Silica: A general term for silicon dioxide.

Siltstone: Similar to shale, but has greater than 2/3 silt and lacks the fine laminations.

Slate: A very fine-grained rock, most often generated from the metamorphism of shale, exhibiting excellent cleavage.

Soapstone: A light-coloured rock with a soft, soapy feel, having a fibrous or flaky texture and composed chiefly of talc.

Syenite: Orthoclase (Potassium) feldspar greater than 2/3 total feldspar, quartz less than 10%.

Trachyte: Extrusive equivalent of syenite.

Travertine: A dense, frequently concentric, form of calcium carbonate created by the rapid chemical precipitation from ground waters (limestone cave formations) or by evaporation around hot springs.

Tufa: A spongy form of calcium carbonate created by evaporation around springs or from a lake surface.

Tuff: A general term for all consolidated pyroclastic material, but typically refers to volcanic ash.

Unconsolidated: Sediment that has not been lithified. Describes the surficial layer below the soil horizons but above bedrock.

- **litter layer**

1. the layer of organic debris, mainly bark, twigs and leaves, on the forest floor.

2. the loose, relatively undecomposed organic debris on the surface of the forest floor made up typically of leaves, bark, small branches and other fallen material.

- **littoral zone**
the shore zone between the high and low water mark.

- **living modified organism**
living organism that has new genetic material acquired through the application of biotechnology.

- **load**
1. to pick up logs and to place them on a vehicle for transportation.
2. the object(s) being hauled.

- **load path**
the arc traced by a turn of logs along a skyline and generally plotted graphically to determine ground clearance.

- **loaded deflection**
deflection when the skyline is supporting a load of logs.

- **loader**
any of a variety of machines, wheel or track mounted, designed primarily to lift and load a truck, train or other mode of transportation.

- **loading**
the act of placing material on a vehicle for further transport.

- **loam**
soil containing a mixture of clay, silt, sand and humus.

- **local planning**
a term describing a variety of resource planning initiatives undertaken to develop integrated approaches to resource use and development.

- **log**
a section of a tree bole at least 8 feet long, not containing a fork, sufficiently straight and sound enough to yield at least an 8-foot board.

- **log boom**
floating logs tied together in rafts to be towed by boat to their destination.

- **log brand**
an ownership identification mark stamped into the end of a log.

- **log dump**
a central unloading area for loads of logs.

- **log rule**
any of several formulae, diagram layouts, or tables for estimating the volume of lumber that can be sawed from logs of a specific length and diameter.

- **log scale**
  the lumber contents of a log as determined by a log rule.

- **log train**
  a log hauling truck designed to haul only short logs; logs usually less than 20 feet long. The normal log truck needs bunk and side logs of 32 or more feet long to bridge from the front to the rear bunks.

- **logger**
  1. a lumberjack; a person whose work is logging.
  2. a worker employed in the woods producing wood products or in the support of such production, such as road construction.

- **logging**
  any or all part of converting trees into logs and transporting them to an unloading area.

- **logging (cutting) plan**
  a map, along with a written plan, describing the road building, harvesting, and other related operations that are submitted for a forest officer's approval to ensure that the applicable standards and obligations stated in the Pre-Harvest Silviculture Prescription and the harvesting agreement are met.

- **logging debris**
  that unwanted, unutilised and generally unmarketable accumulation of woody material in the woodland such as limbs, tops, cull logs and stumps, that remain as residue after timber harvesting.

- **logging trail**
  a narrow, temporary path used by harvesting equipment.

- **logging truck**
  a special 18-wheeled truck designed to haul logs from the woods to the mill.

- **logotherapy**
  logos is a Greek word that denotes 'meaning.' According to logotherapy, striving to find meaning is the primary motivational force in man.

- **long butt**
  a swelled or cull portion cut off of a butt log.

- **long corner**
  1. the distance from the landing to the farthest point of the setting.

2. a small patch of timber in a setting that is difficult to yard or is unreachable.

- **long log**
  to produce logs over 20 feet long , commonly 32 to 40 feet long. AKA 'long wood'.

- **Long Run Sustainable Yield (LRSY)**
  the long run sustainable yield for any Timber Supply Area (TSA) is equal to the culmination of mean annual increment weighted by area for all productive and utilisable forest land types in that TSA including all not satisfactorily restocked, disturbed stocking doubtful, and potentially usable non-commercial cover.

- **longwood**
  1. pulpwood 120 inches or more in length.
  2. stemwood delivered in lengths that exceed 15 feet.

- **longwood harvesting**
  timber harvesting method in which harvested trees are moved to the landing either as whole trees or as topped and limbed tree-length logs. At the landing, further processing such as limbing, topping, bucking, chipping or loading is carried out as necessary.

- **looper**
  a caterpillar in which some or all of the middle abdominal prolegs are wanting and which moves by placing the posterior part of the abdomen next to the thorax, forming a loop of the intervening segments, then extending the anterior part of the body forward.

- **lopping**
  chopping branches, tops and small trees after felling into lengths such that the resultant slash will lie close to the ground.

- **lopping and scattering**
  lopping the slash created after felling and spreading it more or less evenly over the ground without burning.

- **lorica**
  an organic or inorganic casing or shell incompletely surrounding an organism, usually loose fitting.

- **loss factors**
  reductions made to gross timber volumes to allow for decay, waste and breakage.

- **Low Ground Pressure (LGP) machines**
  machines that exert a total ground pressure of less than 43.4 KPa (6.3 pounds per square inch).
- **lowboy**
  a heavy equipment moving truck capable of transporting logging and earthmoving equipment distances greater than which they would move themselves.
- **lumber mill**
  a sawmill that produces lumber.
- **lumberjack**
  a person whose work is cutting down timber and preparing it for the sawmill.
- **machine rate**
  cost per unit of time for owning and operating a logging machine or some other piece of logging equipment. In accordance with engineering practices, the rate is composed of fixed costs such as depreciation, interest, taxes, and license fee, and variable costs including fuel, lubricants etc.
- **macronucleus**
  one of two types of nuclei found in ciliates. Typically the larger of the two, may be rounded, like a long sausage, or like a string of beads. Involved in production of proteins but not in sexual reproduction. Essential for the day to day activities of the ciliate.
- **magnet schools**
  in contrast to comprehensive neighbourhood schools, Magnet Schools are organised by academic or vocational theme and are open to all comers.
- **main road**
  road that supports a high level of traffic, usually well built and well designed.
- **mainline**
  the cable used to haul logs into the landing. Some cable yarder configurations, such as running skyline, may have more than one mainline.
- **mainline block**
  the block on the spar or tower through which the mainline runs.

▪ **major culvert**
a stream culvert having a pipe diameter of 2000 mm or greater, or a maximum design discharge of 6 m^3/sec or greater.

▪ **managed forest**
any forestland that is treated with silvicultural practices and/or harvested.

▪ **managed forest land**
forest land that is being managed under a forest management plan utilising the science of forestry.

▪ **managed harvest**
estimated volume of timber on commercial forest land that could be cut annually for the next 10 years while improving tree stocking and bringing about a more even distribution of age classes. Annual managed harvest is considered separate from harvest cuttings and thinnings and is determined by computer using an area control system that specifies the number of acres to be cut annually.

▪ **management**
see **forest management**.

▪ **management area**
stands or forest types that require similar management practices and can be grouped for treatment as a management unit.

▪ **management assumptions**
approximations of management objectives, priorities, constraints and other conditions needed to represent forest management actions in a forest planning model.

▪ **management option**
a prescription of management activities over time that will achieve specified management objectives.

▪ **management plan**
1. a management plan or management and working plan approved under a tree farm licence, woodlot licence, pulpwood agreement or forest licence. Contains inventory and other resource data.
2. detailed long-term plan for a forested area. Contains inventory and other resource data.
3. a document prepared by natural resource professional to guide and direst

the use and management of a forest property.
4. It consists of inventory data and prescribed activities designed to meet ownership objectives.

- **management zone**
the outer portion of a riparian management area situated adjacent to a stream, lake, or wetland and established to conserve and maintain the productivity of aquatic and riparian ecosystems when harvesting is permitted.

- **mann, horace**
educator, considered the father of American public education, who saw a place where the sons and daughters of lawyers and doctors would mix with the children of farmers and bricklayers. As administrator of the Massachusetts public schools system, he made reforms that served as a model for the entire nation.

- **mantle**
the layer of the earth between the crust and core.

- **map folio**
a series of maps bound together, often produced as overlays of information, e.g., soils, fish, water, forest and wildlife.

- **marginal land**
land that does not consistently produce a profitable crop because of infertility, drought or other physical limitations such as shallow soils.

- **marginal log**
the smallest log, in volume, that it pays a logger to remove from the woods. Often thought of as the breakeven log.

- **marine**
pertaining to the sea.

- **mark**
to select and indicate (usually with paint) trees to be felled in a logging operation.

- **market pulp**
white pulp from hardwood trees that is dried into thick blotter-like sheets and baled for shipment to a paper mill for repulping to make paper products.

- **marketing**
the selling of timber or other forest resources. Successful sellers seek a satisfactory

price through competition, skillful negotiation, knowledge of timber markets and the aid of a competent broker or consultant.

- **marking**

1. the physical process of selecting trees to be cut or left during a harvest.
2. delineating a boundary. Marking is usually done by spraying a spot of bright paint on a prominent part of the tree.
3. selecting and indicating, by a blaze or paint sport, the trees to be cut or left in a timber cutting operation.

- **mass movement**

the downslope movement of earth caused by gravity. Includes but is not limited to landslides, rock falls, debris avalanches and creep. It does not however, include surface erosion by running water. It may be caused by natural erosional processes, or by natural disturbances (e.g., earthquakes or fire events) or human disturbances (e.g., mining or road construction).

- **mass wasting**

movement of soil and surface materials by gravity.

- **mast**

1. fruits or nuts used as a food source by wildlife. Soft mast include most fruits with fleshy coverings, such as persimmon, dogwood seed or black gum seed. Hard mast refers to nuts such as acorns and beech, pecan and hickory nuts.
2. a word used to describe general food sources for wildlife. Acorns, hickory nuts, walnuts and pecans are some examples of hard mast. Blackberries, grapes, persimmons and cherries are some examples of soft mast.
3. all fruits of trees and shrubs used as food for wildlife. hard mast includes nutlike fruits such as acorns, beechnuts and hickories, etc. Soft masts includes the fleshy fruits of black cherry, dogwood and serviceberry, etc.

- **mastigophore**

projection of a cell that supports one or more flagella.

- **mat**

temporary roadway constructed of hardwood lumber.

mature
trees or stands that are sufficiently developed to be harvestable.

mature timber
1. stands of timber where the age of the leading species in a stand is greater than the specified cutting age. Cutting ages are established to meet forest management objectives. Usually stands with lodgepole pine or a deciduous species as the leading species are classified as mature timber when the stand age is greater than 80 years. Otherwise, all stands having conifers other than lodgepole pine and whitebark pine as the leading species are mature when the stand age is greater than 120 years.
2. stand of trees that has attained an age or size that satisfies the primary economic goal for which it was managed.
3. a tree that has reached a desired size or age for its intended use. Size, age, or economic maturity varies depending on the species and intended use.

maturity
see **economic maturity** and **biological maturity**.

maximum density
the maximum allowable stand density above which stands must be spaced to a target density of well-spaced acceptable stems to achieve free-growing status.

MBF
1. thousand board feet.
2. abbreviation denoting 1,000 board feet. MBF is a typical unit of trade for dimension lumber and sawtimber stumpage. (It takes 11 MBF of wood to build an average 1,900-square-foot house.)
3. abbreviation for 1,000 board feet; the unit of trade for lumber and stumpage.

MDF-medium density fibreboard
a composite board, made of wood fibres, that has a smooth surface and good interior machinability. Used for furniture, cabinets, mouldings and picture frames.

Mean Annual Increment (MAI)
1. total increment growth up to a given age divided by

that age. Average growth per year.

2. the average annual increase in volume of individual trees or stands up to the specified point in time. The MAI changes with different growth phases in a tree's life, being highest in the middle years and then slowly decreasing with age. The point at which the MAI peaks is commonly used to identify the biological maturity of the stand and its readiness for harvesting.

- **meandering line**
  1. survey line at the high water mark on navigable lakes and streams.
  2. line at which continuous vegetation ends and sandy or muddy shore begins.

- **mechanical harvesting**
  cutting with mechanised equipment, such as the carrier mounted shear or fellerbuncher, instead of by hand with a power saw.

- **mechanical site preparation**
  any activity that involves the use of mechanical machinery to prepare a site for reforestation.

- **mechanised**
  harvesting and processing timber with machines, reducing manual labour on the ground.

- **mechanised access and use**
  refers to access and use by, for example, mountain bikes and other bicycles, hang gliders, and other human-powered mechanised equipment. Associated facilities include aircraft landing facilities, boat docks and heliports.

- **mechanised logging**
  logging in which most or all of the hand labour is replaced by machines; requires a large outlay of capital.

- **mechanised stand tending treatment**
  any stand tending activity that involves the use of mechanical machinery to treat a stand.

- **membranelle**
  a compound structure comprised of many cilia and associated with the mouth of a ciliate. Either present in groups of three (oligohymenophora) or as a band of many more (polyhymenophora).

- **Memorandum Of Understanding (MOU)**
  an agreement between ministers defining the roles and responsibilities of each ministry in relation to the other or others with respect to an issue over which the ministers have concurrent jurisdiction.

- **mensuration**
  1. in forestry, the measurement of both standing and harvested timber.
  2. the measurement and calculation of volume, growth and development of individual trees or stands and their timber products.
  3. a measurement of forestlands.

- **merchantable**
  logs exceeding a minimum size and a minimum usable volume that are suitable for sale.

- **merchantable**
  logs exceeding a minimum size and a minimum usable volume that are suitable for sale.

- **merchantable height**
  1. the stem length, normally measured from the ground to a 10-, 6-, or 4-inch diameter top, above which no other saleable product can be cut. Diameter, local markets, limbs, knots and other defects collectively influence merchantable height.
  2. the tree height, normally to a 10, 8, 6, or 4 inch diameter top, beyond which no other marketable product can be cut. Local markets and defects influence merchantable height.

- **merchantable timber**
  a tree or stand that has attained sufficient size, quality and/or volume to make it suitable for harvesting.

- **merchantable top**
  smallest utilisable top.

- **merchantable volume**
  the amount of sound wood in a single tree or stand that is suitable for marketing under given economic conditions.

- **meridian line**
  a north-south reference line often appearing on maps. Meridian lines are also etched into the bearing plate on a compass.

- **metabolite**
product of a chemical reaction that required the breaking down of a molecule by an enzyme in order for the cell to obtain energy for vital functions.

- **metric tons**
Unit of weight equal to 1,000 kilograms. Approximately 2,205 pounds.

- **microclimate**
generally the climate of small areas, especially insofar as this differs significantly from the general climate of the region. Stands often create microclimates.

- **micronucleus**
one of two kinds of nuclei found in ciliates, dividing to produce two similar nuclei during asexual reproduction and producing nuclei with half the complement of

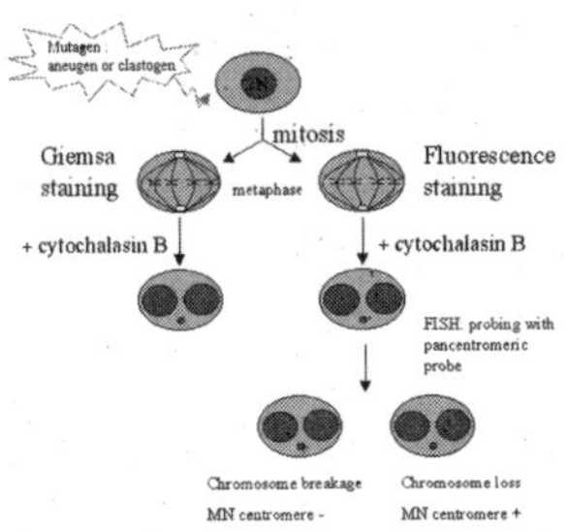

DNA for sexual activity. Usually the smaller of the two types of nuclei, but many may be present. some cells lack a micronucleus and survive quite well without it.

- **microorganism**
1. usually single-celled creature that is too small to be seen without the aid of a microscope. The term may be applied to bacteria, archaea, protists and viruses.
2. a general term for a unicellular or multicellular microscopic organism. Classifications of microorganisms include algae, bacteria, fungi, protozoa and viruses.

- **micropropagation**
an in vitro asexual plant reproduction method. This technique creates progenies that are genetically identical to the parent. The process is analogous to photocopying for creating identical multiple copies.

- **microsite**
a small area which exhibits localised characteristics different from the surrounding area. For example, the microsites created by a rock outcrop with thin soils, or the shaded and cooled areas

created on a site by the presence of slash.

- **microtubule**
a subcellular structure comprised of the protein tubulin and used for support - part of the cytoskeleton. Individual microtubules cannot be seen by conventional light-microscopy, but aggregates of microtubules can.

- **middorsal**
in the middle of the upper side or dorsum.

- **mill residues**
wood materials (coarse and fine) and bark generated at manufacturing plants (primary wood-using mills) when roundwood products are processed into primary wood products, includes slabs, edgings, trimmings, sawdust, veneer clippings and cores, and pulp screenings. (Note: Includes mill residues recycled as by-products as well as those left unutilised and disposed of as waste.)

- **milt**
salmon sperm.

- **mimicking symptoms**
symptoms similar to those caused by pollutants but induced by other abiotic or biotic causal agents.

- **miner**
the larval stage of an insect which makes galleries or burrows between the upper and lower surfaces of leaf tissue.

- **mineral commodities**
a mineral resource with sufficient value to be individually produced and traded on the open market.

- **mineral resource**
a known or undiscovered concentration of naturally occurring solid, liquid, or gaseous material in or on the Earth's crust in such form and amount that economic extraction of a commodity from the concentration is currently or potentially feasible. The following are the various types of mineral resources:
Identified, Demonstrated: An in-place deposit for which there are physical measurements of size and grade combined with limited extrapolation of geological, geochemical and geophysical data.
Identified, Inferred: A known deposit for which there are

only general quantitative estimates of size and grade. The estimates are based on assumed data from geologic evidence and related mining experience
Undiscovered, High Potential: As yet undiscovered deposit in a region where the geologic controls necessary for the formation of deposits of a specific type are present and favourable and evidence indicates mineral accumulation.
Undiscovered, Low Potential: As yet undiscovered deposit in a region where the geologic, geochemical, and geophysical characteristics do not indicate a favourable environment or where there is little evidence that the geologic processes necessary for mineralisation have occurred.
Undiscovered, Moderate Potential: As yet undiscovered deposit in a region where the geologic controls necessary for the formation of deposits of a specific type are favourable and there is a reasonable chance for mineral accumulation.
Undiscovered, Unknown Potential: A region where the geologic information is insufficient to otherwise categorise potential.

- **mineral soil**

soil consisting predominately of, and having its properties determined by, inorganic matter. Usually contains less than 20 percent organic matter.

- **minimum utilisation standard**

included in every licence authorising the harvesting of timber, a standard which is expressed as a maximum stump height, diameter at stump height, and top diameter and which can vary by species and timber supply area (and supply blocks within timber supply areas).

- **mistletoe infection rating**

the relative abundance of mistletoe in the crown of a tree or shrub.

- **mite(s)**

small, often minute, arthropods in the order Acarina of the class Arachnida, which includes spiders, scorpions, and related forms. Mites have four pairs of legs vs. three pairs in insects.

mitochondrion
site of aerobic respiration (energy production that is dependent on oxygen) in eukaryotic cells.

mixed stand
1. a stand composed of two or more tree species.
2. a timber stand in which less than 80 percent of the trees in the main canopy are of a single species.
3. a stand where less than 80% of the trees in the main canopy are of the same species.

mixed wood forest
a forest with softwood and hardwood trees.

mixing height
the height above the surface (m) defining a boundary layer within which pollutants are free to mix.

mixotrophic
used in reference to organisms which use a mixture of nutritional strategies, such as organisms with chloroplasts and carry out photosynthesis but are also able to feed by phagocytosis.

MMBF
Million Board Feet.

mobile tailhold
a crawler tractor or hydraulic excavator fitted with a fair lead or short spar used to hang tailblocks which simplifies and speeds up road changes.

model
theoretical abstraction, usually capable of mathematical manipulation, used to evaluate a problem or a subject of interest.

modern biotechnology
techniques developed that are rapid, efficient, and precise in the manipulation of living organisms. Most modern biotechnology techniques focus on plants at the genetic level.

modified burning zone
a zone within or adjacent to a smoke-sensitive area that requires special considerations and burning techniques, even under favourable conditions, to maintain air quality within a smoke-sensitive area.

moisture content
1. the weight of water in wood, expressed as a percent of the oven-dried weight of the wood.

2. amount of water present in a material such as wood or soil. Generally expressed as a percentage of the material's oven dry weight.
3. amount of water in a material, expressed as a percentage of the material's total weight; used in the pulp and paper industry.

- **molecular marker**
  a gene or gene product of known location on the DNA strand that is used to mark other genes.

- **molt**
  to cast off the outgrown skin or cuticle in the process of insect development; changing from one instar to the next.

- **monoclonal antibody**
  highly specific purified antibody that was derived from only one clone of cells and recognises only one antigen.

- **monoculture**
  in general, even-aged, single-species forest crops.

- **monophyletic**
  a group of organisms that includes their most recent common ancestor and all of its descendents.

- **mopping up**
  act of making a fire safe after it has be controlled by extinguishing or removing burning material along or near a control line.

- **mortality**
  1. death or destruction of forest trees as a result of competition, disease, insect damage, drought, wind, fire and other factors (excluding harvesting).
  2. number or sound wood volume of healthy trees that have died from natural causes during a specified period.

- **mosaic**
  a diseased condition where different portions of a leaf vary in amounts of chlorophyll, thus giving the leaf a mottled appearance; usually caused by viruses.

- **mosses and lichens**
  mosses are plants growing in tufts or clusters on the ground, decaying wood or on rocks. Lichens are any of the various flowerless plants composed of fungi and algae, commonly growing in flat patches on rocks, trees etc.

- **most hazardous pest**
the principal natural agent operating in the vicinity of the sample point which presents the greatest threat to realising stand goals of stocking, growth and structure.

- **motile**
moving, for example by swimming, gliding, crawling, jumping, or kicking. Part of the body (e.g., cilia) may be motile in a cell which is not motile and is fixed in one position.

- **motorised access and use**
refers to access and use by, for example, float planes, helicopters, fixed-wing aircraft, motorboats, motor bikes, all-terrain vehicles, snowmobiles, and motorised equipment.

- **mottle**
irregular, diffuse patterns of chlorotic areas interspersed with normal green leaf tissue.

- **MSR**
machine stress rated lumber is dimension lumber that has been evaluated by mechanical stress-rating equipment. The stress-rating equipment measures the stiffness of the material and sorts it into various modulus of elasticity classes.

- **mucilaginous**
made of, or with the texture of mucus.

- **mucus**
a gelly-like substance produced by organisms, texture may vary from virtually fluid to stiff and rubber-like.

- **mulching**
providing any loose covering for exposed woodland soils, such as grass, straw or wood fibres to protect exposed soil and help control erosion.

- **multicellular**
an organism consisting of more than one cell.

- **multiple entry**
entering a stand for commercial harvesting more than once in any one continuous rotation.

- **multiple use**
1. a system of resource use where the resources in a given land unit serve more than one user. Multiple use can be effected in three ways.

2. different uses of adjacent sub-areas which together form a composite multiple use area; the alternation in time of different uses on the same areas; and more than one use of an area at one time. In multiple use planning, where differing resource uses are conducted at the same time on the same area and conflicts between users will occur, one resource is determined to be the dominant use and all other secondary uses are integrated only in-so-far as they are compatible with the first. Often multiple use planning sacrifices the production of the individual resources in favour of the over-all mix of resource uses that brings the greatest social and economic benefits.
3. the management of land or forest for more than one purpose, such as wood production, water quality, wildlife, recreation, aesthetics or clean air.

- **multiple use and value**
a conceptual basis for managing a forest area to yield more than one use or value simultaneously. Common uses and values include aesthetics, water, wildlife, recreation and timber.

- **multiple-use**
practice of forestry that combines two or more objectives.

- **multiple-use forestry**
concept of forest management that combines two or more objectives, such as production of wood or wood-derivative products, forage and browse for domestic livestock, proper environmental conditions for wildlife, landscape effects, protection against floods and erosion, recreation and protection of water supplies.

- **multiple-use management**
1. management of land resources with the objective of achieving optimum yields of products and services from a given area without impairing the productive capacity of the site.
2. the practice of managing forest resources for a variety of simultaneous uses and benefits including water, forage, wildlife habitat, wood, recreation, wilderness and minerals.

- **multi-purpose forestry**
forest management that delivers multiple benefits e.g. social, economic, environmental

- **multispan skyline**
a skyline having one or more intermediate supports.

- **multistem**
operation handling two or more stems at the same time.

- **multi-track year-round schools**
a plan in which students in one school are divided into three, four, or five groups, each with different schedules of instructional and vacation time. The time students spend in schools is staggered so that one group is on vacation while other groups are in school. Multi-track schools have been created primarily to deal with overcrowding and to make better use of facilities.

- **mutagenesis**
the introduction into a gene of an alteration that changes the structure or function of the gene product.

- **mutagens**
agents that cause a change in the genetic makeup of a cell. These include chemicals, X-rays and ultraviolet light.

- **mutation**
an alteration in the genetic makeup of a cell. These genetic alterations may be unexpressed by the cell.

- **mycelium**
a mass or aggregate of hyphae; vegetative stage of fungi.

- **mycorrhiza**
a rootlet of a higher plant modified through integral association with a fungus to form a constant structure which differs from either component but is attached to the root system and functions somewhat as a rootlet. It is usually considered to be beneficial to the associated plant.

- **naked**
larva devoid of body hairs or setae; pupa not enclosed in a cocoon or other covering.

- **natural boundary**
the visible high water mark of any lake, stream, or other

body of water where the presence and action of the water are so common and usual and so long continued in all ordinary years as to mark upon the soil of the bed of the lake, river stream, or other body of water a character distinct from that of the banks, both in vegetation and in the nature of the soil itself.

- **natural disturbance regimes**
  the historic patterns (frequency and extent) of fire, insects, wind, landslides and other natural processes in an area.

- **natural justice**
  a set of procedures designed to ensure that decisions are made fairly.

- **natural range barrier**
  a river, rock face, dense timber or any other naturally occurring feature that stops or significantly impedes livestock movement to and from an adjacent area.

- **natural regeneration**
  1. the renewal of a forest stand by natural seeding, sprouting, suckering, or layering seeds may be deposited by wind, birds or mammals.
  2. a natural regrowth of seedlings after harvest or after a natural disturbance has occurred, such as a fire.

- **natural resource**
  means land, water and atmosphere, their mineral, vegetable and other components, and includes flora and fauna on or in them.

- **natural stand**
  a stand of trees resulting from a natural seed fall or sprouting.

- **naturally resistant seed sources**
  tree species or provenances that have been shown to exhibit increased resistance to some specific pest, over that of the species or provenance that would normally be used in artificial regeneration in a particular situation.

- **necrosis**
  death.

- **needle leaf**
  a leaf from a conifer.

- **nemadesmata**
  stiff aggregates of many microtubules found around the cytostome of some ciliates and used during the ingestion of food. A type of ingestion rod.

- **neo-tropical birds**
birds that breed in the northern hemisphere during summer months, and winter in tropical regions (e.g., woodthrush or barn swallows). A significant percentage of Massachusetts' breeding birds are neo-tropical migrants.

- **net annual growth**
increase in volume of trees during a specified year. Components of net annual growth include the increment of net volume of trees at the beginning of the specified year that survive to the year s end, plus the net volume of trees reaching the minimum size class during the year, minus the volume of trees that died during the year, and minus the net volume of trees that become rough or rotten trees during the year.

- **net down procedure**
the process of identifying the net land base, which is the number of hectares of forest land which actually contribute to the allowable annual cut. The process involves netting down the TSA gross area to the TSA gross forest area then to the TSA net forest area. Areas and/or volumes are sequentially deleted or reduced from the gross land base for a number of considerations, including: private ownership, non- forest or non-productive, environmentally sensitive, unmerchantible and inaccessible.

- **net land base**
see **net down procedure**.

- **Net Present Value (NPV)**
a stand's present worth before harvesting once costs associated with its establishment and tending have been subtracted.

- **net volume**
volume of the main stem excluding stump and top as well as the defective and decayed wood of trees or stands.

- **neuston**
the environment of the interface between water and air, often rich in bacteria and protists.

- **new forestry**
a philosophy or approach to forest management that has as its basic premise the protection and maintenance of

ecological systems. In new forestry the ecological processes of natural forests are used as a model to guide the design of the managed forest.

- **niche**
  1. the place and function of a species in an ecosystem.
  2. the physical and functional location of an organism within an ecosystem; where a living thing is found and what it does there.

- **node**
  1. that point on a stem which usually bears branches and leaves.
  2. a branching point in a phylogenetic tree, usually representing the common ancestor of the lineages descending from this branching point. In the Tree of Life Project, each node represents this common ancestor and all of its descendents.

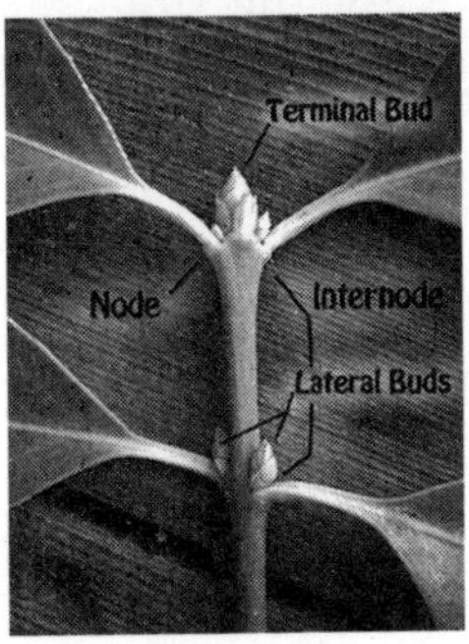

- **non-clamping carriage**
  skyline carriage which is held in place by counter-tension of the mainline(s) and the haulback line, rather other than clamping to the skyline line a clamped carriage.

- **noncommercial species**
  tree species in which small size, poor form or inferior quality is typical. These species do not normally develop into trees suitable for conventional forest products.

- **non-designated wilderness**
  areas within the provincial forest that have been zoned as wilderness through approved integrated resource management plans including regional land-use plans and Land and Resource Management Plans (LRMPs).

- **non-forest land**
  land not primarily intended for growing or supporting a forest.

- **Non-Industrial Private Forestland (NIPF)**
  forestland owned by a private individual, group or corporation not involved in wood processing.

■ **nonpoint source pollution**
1. water pollution created from a source that has no particular or permanent point of input.
2. occurs when rainfall or snowmelt runoff moves across the ground carrying debris, sediment and/or pollutants into streams, lakes, wetlands and groundwater.

■ **nonrenewable resource**
a resource that is in limited supply and doesn't have the capacity to be replaced through natural processes, at least not for many thousands of years. Fossil Fuels are a nonrenewable resource.

■ **non-timber resource values**
values within the forest other than timber which include but are not limited to biological diversity, fisheries, wildlife, minerals, water quality and quantity, recreation and tourism, cultural and heritage values, and wilderness and aesthetic values.

■ **non-timber resources**
resources other than timber, such as recreation, aesthetics, wildlife, fish, forage, range, water and soils.

■ **normal forest**
an outdated concept, drawing on the idea of a norm or standard forest structure against which existing forest structures can be compared. A normal forest is a forest composed of even-aged fully-stocked stands representing a balance of age classes such that for a specified rotation period, one age class can be harvested in each year. At the end of the rotation, the stands that were harvested first in the cycle would be ready for harvesting again.

■ **Not Satisfactorily Restocked (NSR)**
productive forest land that has been denuded and has failed, partially or completely, to regenerate either naturally or by planting or seeding to the specified or desired free growing standards for the site.

■ **notch**
a wedge-shaped piece cut out of a stump to prevent a guyline or block strap from slipping off.

■ **no-work zones**
areas in which equipment and people are not allowed

during forestry operations, usually for safety or ecological reasons.

- **noxious weeds**
any weed so designated by the Weed Control Regulations and identified on a regional district noxious weed control list.

- **nubbin**
a metal knob or collar attached to the end of a choker, allowing it to be secured to a choker hook.

- **nucleated**
having a nucleus.

- **nucleolus**
an optically dense region (or regions) in a nucleus, associated with RNA synthesis. Not always visible.

- **nucleotides**
the building blocks of DNA that are critical for determining the genetic information of an organism. Their sequence determines the structure of the protein(s) made by that cell.

- **nucleus**
1. an organelle found only in eukaryotic cells, in which most of the cellular DNA (genetic material) is located. Most cells have a single nucleus, but certain species may have many nuclei.

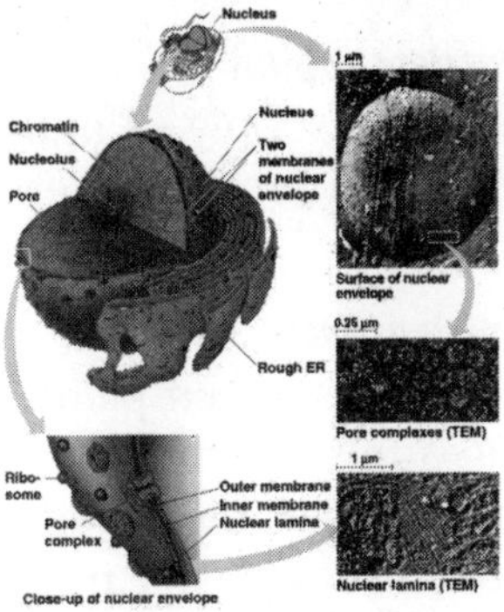

2. a complex spherical body containing chromosomes that are enclosed by a membrane. Nuclei are found in most animal and plant cells.

- **nudipodium**
a type of unsupported pseudopodium without evident extrusomes (compare extrusopodia).

- **nurse log**
a larger and decomposing fallen log which acts as a germination substrate for tree species establishing in the understorey. Such logs provide moisture, nutrients and often some degree of elevation above other potentially competing vegetation on the forest floor.

nursery
plot of ground on which young plants are reared for transplantation.

nutrient capital
the total supply of nutrients in the organic matter and the soil.

nutrient cycle
producers (green plants) use the sun's energy to manufacture their own food from abiotic (non living) elements in a process called photosynthesis. These green plants provide food and oxygen for other living things (consumers). Some consumers (herbivores) eat producers while some (carnivores) eat other consumers. Decomposers break down dead plant and animal materials into abiotic elements. Decomposers are recyclers; the abiotic elements return to the soil, water, and air for use again. This is how nutrients are cycled and recycled.

nutrient
1. substance required for growth and development. Plants, for example, need water and minerals to grow and reproduce.
2. chemical element essential for the growth and survival of an organism.
3. elements necessary for growth and reproduction. Primary plant nutrients are nitrogen, phosphorus and potassium.
4. mineral elements in the forest ecosystem such as nitrogen, phosphorus, or potassium, that are naturally present or may be added to the forest environment by forest practices such as fertiliser or fire retardant applications. Substances necessary for the growth and reproduction of organisms. In water, those substances that promote growth of algae and bacterial; chiefly nitrates and phosphates.
5. in the forest context, mineral elements such as nitrogen, phosphorus, or potassium, that are naturally present or may be added to the woodland environment by woodland practices such as fertiliser or fire retardant applications. Substances neces-

sary for the growth and reproduction of organisms. In water, substances that promote growth of algae and bacteria; chiefly nitrates and phosphates.

- **nymph**

the immature stage of insects, following hatching, which does not have a pupal stage, i.e., incomplete metamorphosis (egg, nymph, adult). Late instar nymphs may have nonfunctional rudimentary wings and/or genitalia.

- **objective**

the end result(s) that must be achieved through management at any given administrative level. Objectives are quantified and indicate time and agency responsibility.

- **Old Corrugated Containers (OCC)**

brown boxes that have been used for their intended purpose, then collected for recycling.

- **odour type and concentration**

the threshold ambient concentration at which certain pollutants are odoriferous to humans.

- **off-highway truck**

a truck designed to handle loads exceeding legal highway size and weight restrictions.

- **offside**

the side opposite that on which the cutter must stand to fell or buck a tree.

- **old growth**

old growth is a forest that contains live and dead trees of various sizes, species, composition, and age class structure. Old-growth forests, as part of a slowly changing but dynamic ecosystem, include climax forests but not sub-climax or mid-seral forests. The age and structure of old growth varies significantly by forest type and from one biogeoclimatic zone to another.

- **old growth forest**

forests composed of diverse stands of mature trees of various species.

- **old-growth**

forests that approximate the structure, composition, and functions of native forests prior to European settlement. They vary by forest type, but generally include more large trees, canopy layers, standing snags, native species and dead organic matter than do young or intensively managed forests.

- **old-growth associated species**
  plant and animal species that exhibit a strong association with old-growth forests.

- **old-growth attributes**
  structural features and other characteristics of old-growth forests, including: large trees for the species and site; wide variation in tree sizes and spacing; accumulations of large dead standing and fallen trees; multiple canopy layers; canopy gaps and understory patchiness; elements of decay such as broken or deformed tops or trunks and root decay; and the presence of species characteristic of old growth.

- **old-growth forest**
  a forest stand usually at least 180-220 years old with moderate to high canopy closure; a multilayered, multispecies canopy dominated by large overstory trees; high incidence of large trees; some with broken tops and other indicators of old and decaying wood (decadence); numerous large snags; and heavy accumulations of wood, including large logs on the ground.

- **old-growth management areas**
  areas which contain, or are managed to replace, specific structural old-growth attributes and which are mapped out and treated as special management areas.

- **omnivore**
  animal organisms that can use plants and other animals as food sources.

- **on the stump**
  standing, uncut timber.

- **ongrowth**
  see **tree history**.

- **open grown or isolated trees**
  trees or shrubs with crowns receiving full light from above and from all sides.

- **opening line**
  a line used to open a grapple.

- **open-side carriage**
  a skyline carriage that opens on one side to enable it to

pass intermediate support jacks.

- **operability line**
a line drawn on a map to differentiate between areas that are operable and those that are not, given status quo harvesting and reforestation technology. Inoperable areas are not economically viable to harvest without seriously impairing the site or other resource values. The operability line is used to determine the operable land base in long-run, sustained yield calculations.

- **operable forest**
that portion of the production forest that, under current market conditions, can be harvested at a profit.

- **operable land**
all lands that are not considered inoperable lands. See **inoperable lands**.

- **operable timber**
see also **timber operability**. Available timber that can be economically logged with present harvesting methods after consideration of access, timber quality and market price.

- **operating area**
geographic sub-units of timber supply areas that have been assigned to individual major licensees for the purposes of long-term planning. The boundaries are subject to change as the timber profile within a timber supply area changes over time.

- **operational cruise**
an estimate, to a specified degree of accuracy, of the volume of timber on an area to be harvested.

- **operational plan**
methods, schedules and responsibilities for accessing, harvesting, renewing and protecting the resource are set out to enable site-specific operations to proceed. Operational plans include forest development plans, logging plans, access management plans, range-use plans, silviculture prescriptions, stand management

prescriptions and five year silviculture plans.

- **opposite**
2 or 3 leaves that are directly across from each other on the same twig.

- **optimum road spacing**
the distance between parallel timber access roads which minimises total logging cost.

- **option**
1. a set of assumptions representing a possible management direction. Options are constructed as a normal part of a planning process in order to provide a framework for analysis and to facilitate management decision-making.
2. the different alternatives available for action. All options have strengths, weaknesses, risks and trade-offs. Informed decisions are made by understanding the options, and then choosing the best one. There are many options for running an organisation or managing a forest ecosystem.

- **ordinary high watermark**
the point on the bank or shore up to which the presence and action of the water is so continuous as to leave a distinct mark either by erosion, destruction of terrestrial vegetation, or other easily recognised characteristics.

- **organelle**
a discrete structure found within eukaryotic cells.

- **organic**
referring to or derived from living organisms; in chemistry, any compound containing carbon.

- **organic matter**
decomposed plant and animal residues.

- **organic soil**
soil containing a high proportion (greater than 20 or 30 percent) of organic matter.

- **orthophovo**
a completely rectified copy of an original photograph. All variations in scale and displacements, due to relief, have been eliminated, hence the name ortho (correct) photography. Orthorphoto and orthophoto map are synonymous, an orthophoto is, very simply, a photo map.

- **Oriented Strand Board (OSB)**
  structural panels made of narrow strands of fibre that are oriented lengthwise and crosswise in layers and bound together with resin.

- **osmotrophic**
  a form of nutrition in which soluble compounds are taken up by the organism, either by pinocytosis or by mechanisms capable of transporting one or a few molecules at a time (membrane pumps).

- **outhaul**
  the portion of a cable yarding cycle where the butt rigging or carriage returns to the timber from the landing for another turn.

- **outslope**
  to shape the road surface to direct water away from the cut slope side of the road.

- **ovary**
  the inner part of carpel or pistil where eggs are borne.

- **overlanding**
  placing road construction fill over organic soil, stumps and other plant materials, corduroy or geotextiles, any of which is required to support the fill.

- **overlay**
  a transparent sheet (clear film material) accompanying a map, on which information, colouring, or symbols are entered so that when the overlay is placed on the map the effect is identical to having entered the overlay information on the map, itself.

- **overmature**
  1. in even-aged management, those trees or stands past the mature stage.
  2. point at which timber has begun to lessen in commercial value because of size, age, decay, or other factors. Many trees in a virgin or old growth stand are overmature and are, in fact, dying of old age.

- **over-mature trees**
  trees that have grown beyond the stage when they would be harvestable. They are declining in growth rate and economic value.

- **overrun**
  difference between the log scale of a shipment of timber and the volume of ac-

tual lumber obtained from it.

- **overstorey**
  1. that portion of the trees in a forest of more than one storey forming the upper or uppermost canopy layer.
  2. layer of foliage in a forest canopy including the trees in a timber stand. Tall mature trees that rise above the shorter immature understory trees.
  3. trees that provide the uppermost layer of foliage in a forest with more that one roughly horizontal layer of foliage.

- **overstory removal**
  any silvicultural treatment with the desired end result being the removal of the overstory component from the growing stock of a multistoried stand. Examples are outright harvest, girdling and simply felling the overstory.

- **overtopped**
  trees with crowns entirely below the general level of the crown cover receiving little or no direct light from above or from the sides.

- **overtopped trees**
  understory trees or shrubs generally of tolerant species such as Engelmann spruce. If they can be released through some kind of treatment and have not been too badly deformed, such trees will make suitable crop trees. Overtopped trees are often of a younger age than the main canopy.

- **overtopping**
  vegetation higher than the favoured species, as in brush or deciduous species shading and suppressing more desirable coniferous trees.

- **oviposition**
  the act of depositing the eggs.

- **ovipositor**
  the egg-laying apparatus; the extended genitalia of a female insect.

- **ownership**
  the identification of the legal owner/administrator (Federal, State, Local, Private) on both the surface and subsurface estates.

- **ozone**
  an unstable, pale-blue gas with a penetrating odour.

- **ozone layer**
layer of gaseous ozone in the stratospheric atmosphere that protects life on earth by filtering out harmful ultraviolet radiation from the sun.

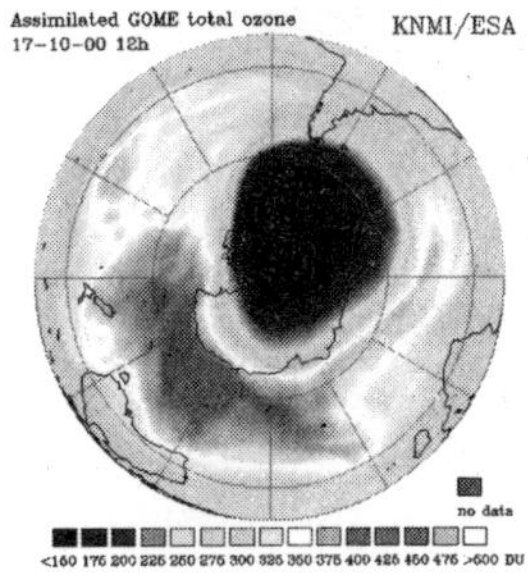

- **pacific salmon**
members of the Salmonidae family and Oncorhynchus genus including; Chinook, Coho, Sockeye, Pink and Chum. Also includes sea-run Steelhead and Cutthroat.

- **paleontological resources**
any remains, trace, or imprint of a plant or animal that has been preserved in the Earth's crust since some past geologic time. Listed below are some of such resources:
Invertebrate Fossil: Remains, including casts and molds, of animals without backbones and not classed as a microfossil.
Microfossil: Fossilised remains of organisms that are too small to be studied without the aid of a microscope.
Petrified Wood: Material formed by the replacement of the wood by silica in such a manner that the original form and structure is preserved.
Plant: Members of the plant kingdom excluding petrified wood and microfossils.
Trace: Indirect evidence of life, such as tracks, trails, burrows and coprolites.
Vertebrate Fossil: Remains, including casts and moulds, of animals with skulls and internal skeletons.

- **palmate**
having leaflets radiating out from a central point.

- **paper plan**
an initial timber harvest plan using available maps, aerial photos and timber inventory data. It serves as the basis for field verification and adjustment to the final logging plan.

- **paradigm**
this is a Greek word for a model or a set of rules and regulations which set

boundaries and tell you how to be successful within those boundaries. A mental model. We have paradigms for everything. Organi-sations are forests of paradigms. Leaders are capable of moving people from one paradigm to another.

- **paradigm effect**

  when ingrained paradigms filter incoming information, blinding people to the presence of new and better models. If the paradigm effect is strong enough it becomes 'paradigm paralysis.'

- **paradigm shift**

  what happens when a new model comes along rendering the old model obsolete. In every shift there are winners and losers.

- **paraphyletic**

  a group of organisms that includes their most recent common ancestor and some but not all of its descendents. Technically speaking, the distinction between polyphyletic and paraphyletic is not as clear as it might seem.

- **parasite**

  1. an organism which lives on or in another living organism and obtains part or all of its nutrients from that other living organism.
  2. an organism that lives on or in an organism of another species and derives nutrients from it.
  3. an organism that lives in or on the living tissue of another organism (the host), deriving nutrients at the expense of the host. Parasites usually live for some time in association with their hosts rather killing them soon after encounter.

- **paraxial rod**

  a rod of material lying within the flagellum parallel to the axoneme, only found in some protists (e.g. euglenids) causing the flagellum to appear relatively thick.

- **parent material**

  parent material is the unconsolidated organic and mineral material in which soil forms. Some soils have formed from the weathering of bedrock in place; however, other soils have formed from material that has been transported from the site of the parent rock and redeposited at a different location through the action of glacial ice, water, wind or gravity.

- **parr**

also known as a fingerling. A large juvenile salmonid, and between a fry and a smolt.

- **partial cutting**

refers generically to stand entries, under any of the several silvicultural systems, to cut selected trees and leave desirable trees for various stand objectives. Partial cutting includes harvest methods used for seed tree, shelter wood, selection and clearcutting with reserves systems.

- **partial cutting**

1. trees removed other than by clearcutting.
2. any harvest in which not all of the trees are removed.

- **particle gun transformation**

a method of genetic engineering that uses a device to accelerate particles of DNA into a cell. The desired DNA sequence is incorporated into the cell's original DNA resulting in a transgenic species.

- **particle size**

the effective diameter of a particle measured by sedimentation, sieving, or micrometric methods. Particle size distribution is the percent by weight of sand, silt and clay in a soil sample of a soil horizon excluding coarse fragments, where:
Coarse Fragments: Rock or mineral particles >2.0 mm in diameter.
Sand: A soil particle between 0.05 and 2.00 mm in diameter. Any one of five soil separates namely: very coarse sand; coarse sand; medium sand; fine sand; and very fine sand.
Silt: A soil separate consisting of particles between 0.05 and 0.002 mm in equivalent diameter.
Clay: A soil separate consisting of particles <0.002 mm in equivalent diameter.

- **pass**

in timber harvesting, one of a planned sequence of harvesting operations designed to harvest a management unit over an extended period of time in discrete phases, so that the size of individual cutblocks and the total area harvested in any one pass does not exceed prescribed limits.

- **patch**

1. in landscape ecology, a particular unit with identifiable boundaries which differs from its surroundings in one or more ways. These can be a function of vegetative composition, structure, age or some combination of the three.

2. a small area of a particular ecological community surrounded by distinctly different ecological communities, such as a forest stand surrounded by agricultural lands or a small opening surrounded by forestland.

- **patch cutting**

a silvicultural system that creates openings less than 1 hectare in size and is designed to manage each opening as a distinct even-aged opening.

- **patch dynamics**

the process of recoloni-sation by plant and wildlife species following the creation of a patch. Small patches and ones close to a source of plant and animal species will be recolonise faster than larger, more isolated patches.

- **pathogen**

1. any agent of the environment capable of inciting disease.

2. a disease-causing organism.

- **pathological rotation age**

the maximum rotation age through which a stand of trees may be grown without significant volume loss from disease. The stand age at which annual volume loss from disease equals annual volume increment.

- **pay-as-cut**

timber purchase based on a dollar amount for a certain amount/volume of wood, such as dollars per MBF. Payment is made only as timber is cut and transported.

- **peaker**

the top log on a loaded logging truck.

- **peavey**

1. stout wooden lever, fitted with a strong, sharp spike used for rolling logs.

2. a short or long pole with a metal spike and small hook on one end, used to turn and manoeuvre logs.

- **pecker pole**
  a small log.

- **pedicel (pedicuncle)**
  the flower stem.

- **peeler**
  high-grade log from which veneer is peeled, on a lathe, for the production of plywood. Peelers are most frequently from old-growth trees. The resulting veneer is usually clear and used in sanded plywood.

- **peewee**
  a small merchanantable log.

- **pelagic**
  pertaining to the water column of the open sea - as opposed to coastal areas or the sea floor.

- **pellet**
  a little ball made of compressed wood; used for burning in some wood stoves.

- **pellicle**
  the outer region of cytoplasm of some protozoa, the term is applied only when the region can be distinguished because it appears to be relatively stiff and highly structured. Used mostly in reference to euglenids and ciliates.

- **perennial**
  plants that live or grow for more than one year. Some resprout from a root system or reseed themselves every year.

- **perennial stream**
  1. a stream that flows year around.
  2. a stream that maintains water in its channel a majority of the year.

- **perennial wildlife mixture**
  a mixture of all or some of the following: shrub lespedeza, partridge pea, cowpea, annual lespedeza, reseeding soybeans and other perennial plants that are beneficial to wildlife.

- **periodic annual increment**
  mean annual growth or increase in volume during a specific period of time.

- **periodic harvest (periodic cut)**
  the removal of several years' accumulated AAC in one year or other period.

- **periplast**
  the entire assemblage of scales, spines and spicules which encase some heliozoa, chrysomonads, etc.

- **peristalsis**
  regular contractions of a body or part of a body. Mostly said of the intestinal system of vertebrates but also the squirming behaviour of some euglenids.

- **peristome**
  the region of the body around, and external to, the mouth. To deserve application of this term, the region must be modified to favour the acquisition of food.

- **perithecium**
  a sexual fruiting structure of the ascomycetes with an opening called the ostiole at or near its top.

- **permanent access structure**
  a structure, including a road, bridge, landing, gravel pit or other similar structure, that provides access for timber harvesting, and is shown expressly or by necessary implication on a forest development plan, access management plan, logging plan, road permit or silviculture prescription as remaining operational after timber harvesting activities on the area are complete.

- **permanent all-season woodland roads**
  these roads usually have gravel surfaces and are designed for year-round use. However, there may be some restrictions on use at various times of the year.

- **permanent bridge**
  a bridge having all its major components constructed of steel, concrete or pressure-treated timber.

- **perpendicular felling**
  timber felled at right angles to ground contour lines.

- **persistence**
  refusing to give up, especially when faced with opposition or difficulty; continuing firmly or steadily.

- **personal**
  having to do with the character, personality, intimate affairs, conduct, etc. of a certain person .

- **pest**
  1. any forest health agent designated as detrimental to

effective resource management.

2. any organism that is out of place or causes stress to a desired organism.

3. an organism capable of causing damage. Forest pests include insects and diseases.

- **pest incidence**

a measurement of the presence and magnitude of pests within a given area.

- **pesticide**

1. any substance or mixture of substances (other than a device) intended for killing, controlling, or managing insects, rodents, fungi, weeds, and other forms of plant or animal life that are considered to be pests.

2. any chemical used to kill or control pests.

3. any chemical used for killing insects, weeds, etc.

4. any agent or agents whose application is intended to deter, kill, or control a pest.

- **pesticide buffer zone**

a strip of land between the 10 m pesticide-free zone and the pesticide treatment area for preventing entry of pesticides or pesticide residues by drift, runoff, or leachate into the pesticide-free zone.

- **petals**

outer part of a flower; protects the inside parts of a flower.

- **petiole**

thin section of leafstalk joining the base of the leaf to the lamina; generally is cylindrical or semicircular in form. Also called the stalk.

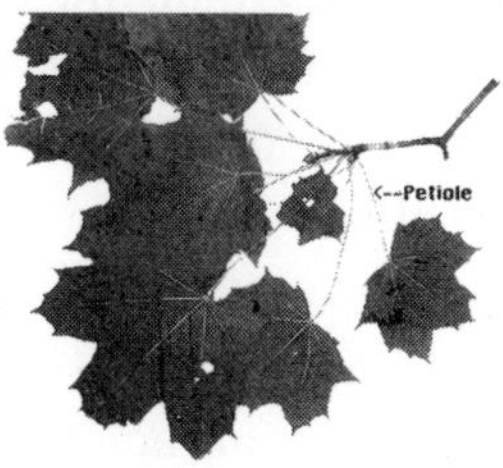

- **pH**

1. a measure of hydrogen ion activity or concentration. Expressed as pH units.

2. the hydrogen-ion activity used in expressing both acidity and alkalinity on a scale whose values range

from 0 to 14, with 7 representing neutrality. Numbers less than 7 represent increasing acidity, and numbers greater than 7, increasing alkalinity.

- **phagocytosis**
  the ingestion of visible particles of food by enclosing them with a membrane to form a food vacuole.

- **phagotroph**
  an organism which feeds by phagocytosis.

- **pharynx**
  a region of the ingestion apparatus lying internal to the mouth (of a metazoan organism) or internal to the cytostome of a protist. Involved in the swallowing process. See **cytopharynx**.

- **phenotype**
  1. an organism as observed by its visible characteristics, resulting from the interaction of its genotype with the environment.
  2. outward appearance or physical attributes of an organism resulting from both the effects of the environment and genetic makeup.

- **phloem**
  1. a layer of tree tissue just inside the bark that conducts food from the leaves to the stem and roots. See **cambium**.
  2. a layer of plant tissue just inside the bark that transports food (dissolved nutrients) from the leaves to the stem and roots.
  3. a spongy layer which stores food manufactured in leaves and transports it to other parts of the tree (inner bark).
  4. a protective layer made up of tiny tubes that transport the sugars from the leaves to the rest of the tree.

- **phosphate**
  a chemical compound that aids root growth and is essential in energy transfer. It is commonly incorporated into beds as Triple Super Phosphate (TSP) at time of planting.

- **photosynthates**
  the products of photosynthesis.

- **photosynthesis**
  1. complex process that takes place in cells of green plants. Radiant energy from

the sun is used to combine carbon dioxide (CO_2) and water (H_2O) to produce oxygen (O) and carbohydrates (such as glucose) and other nutrient molecules.
2. a process by which green plants use the energy from the sun to fabricate their own food.
3. the process by which plants make sugar from sunlight, water and carbon dioxide.
4. a means of acquiring energy for metabolism which involves trapping radiant energy in chloroplasts, the use of that energy to break up water molecules (hydrolysis) and to convert released energy into an accessible form - such as the molecule ATP The only form of autotrophy in eukaryotic cells. Some heterotrophic protists have symbiotic algae which allow them to exploit photosynthesis.

- **photosynthetic pigments**
large molecules in chloroplasts that absorb radiant energy (hence they have colour), mostly chlorophylls and carotenes and occasionally phycobilins.

- **phylogeny**
the evolutionary history of a group of organisms. Any group of species are descended from a common ancestral species, which, over time, split into two species, with these descendents splitting again, and so on, until the entire collection of species was produced. The genealogical connections thereby formed in general take the shape of a tree, called the phylogeny of the group.

- **pinnate**
having leaflets on both sides of a common axis.

- **pinocytosis**
a process of ingesting material by enclosing it with a membrane. The resulting structure is usually too small to be seen with the light microscope and is mostly suitable for the ingestion of fluid or mucus.

- **pioneer**
to do initial work over rough or unclear areas or to make a pioneer road.

- **pioneer plants**
a succession term for plants capable of invading

bare sites, such as a newly exposed soil surface, and persisting there, i.e., 'colonising' until supplanted by invader or other succession species.

- **pioneer road**
  a primitive tractor-trail type of access for moving equipment, workers and material.

- **pistil**
  female part of a flower; also known as the carpel.

- **pitch tubes**
  a tubular mass of resin that forms on the surface of bark at bark-beetle entrance holes.

- **planimeter**
  instrument used to mechanically measure an area by tracing the perimeter on a plane surface. Typically used to estimate the size of an area from scale maps.

- **planing**
  to make smooth or level with a plane.

- **plankton**
  organisms living in the water column (above the sediment).

- **planned grazing system**
  a system approved by the regional manager or district manager respecting the use of land for grazing and the dispersal of livestock over land.

- **planning**
  the determination of the goals and objectives of an enterprise and the selection, through a systematic consideration of alternatives, of the policies, programs and procedures for achieving them. An activity devoted to clearly identifying, defining and determining courses of action, before their initiation, necessary to achieve predetermined goals and objectives.

- **planning horizon**
  the time period which will be considered in the planning process.

- **planning term**
  the term of the actual plan before it must be updated.

- **plant community**
  an assemblage of plants occurring together at any point in time, thus designating no particular ecological status.

- **plant harvesting**
  the collection of plant life including, but not limited to, bark, berries, boughs, branches, burls, cones, conks, ferns, flowers, grasses, herbs, fungi, lichens, mosses, mushrooms, roots, sedges, shrubs, sprays and twigs.

- **plant or habitat diversity**
  a variety of food or cover for wildlife. Variation may occur at one point in time or over a period of time such as during the course of a season. Seasonal diversity of food and cover is often critical to the survival of a species.

- **plantation**
  1. planted pines or hardwoods, typically in an ordered configuration such as equally spaced rows.
  2. a human-made forest usually established by planting seedlings.
  3. forests or woods that have been planted.

- **plant bodies**
  antibodies that are part of an immunological response made by plants to control plant disease.

- **planting**
  establishing a forest by setting out seedlings, transplants or cuttings in an area.

- **Plants With Novel Traits (PNTS)**
  the product of any genetic modification technique that produces a plant with a characteristic previously unseen in its species.

- **plasmid**
  a small circle of bacterial DNA that is used as a vector in the transfer of genes from one organism to another. Plasmids have the ability to replicate independently within a host.

- **plasmodium**
  a type of amoeboid organisation involving a large mass of cytoplasm and, usually, many nuclei. A type of body form adopted by some slime moulds. The genus Plasmodium is the cause of malaria.

- **plastid**
  the same as chloroplast, from it come the terms aplastidic and plastidic for with and without chloroplasts respectively.

- **plesiomorphy**
  an ancestral character state. A trait from which an evolutionary novelty (apomorphy) is derived. Also see **symplesiomorphy**.

- **plot**
  a carefully measured area laid out for experimentation or measurement.

- **plug**
  a seedling grown in a small container under carefully controlled (nursery) conditions. When seedlings are removed from containers for planting, the nursery soil remains bound up in their roots. See **Bareroot seedling**.

- **plus tree**
  a phenotype judged (but not proven by test) to be unusually superior in some quality or qualities such as an exceptional growth rate relative to the site, desirable growth habit, high wood quality, exceptional apparent resistance to disease and insect attack or to other adverse locality factors.

- **point sampling**
  a method of selecting trees for measurements and of estimating stand basal area at a sample location or point sample. Also called plotless cruising, angle count method, Bitterlich method. A 360 degree sweep is made with an angle gauge about a fixed point and the stems with breast height diameters appearing larger than the fixed angle subtended by the angle gauge are included in the sample.

- **pole**
  young tree at least 4 inches and less than 8 to 12 inches in d.b.h. Any considerable length of round timber below saw log size, ready for use after removal of the bark without further conversion. Suitable for power poles or for simple building work. Cedar or red fir log of saw log size, according to an individual pole manufacturer s specifications for length.

- **pole size**
  trees from 5 to 11 inches in d.b.h.

- **pole stand**
  a stand of trees with dbh ranging from 5 to 9 inches.

- **poles or poletimber**
  1. trees from 5 to 7 inches in diameter at breast height.

2. arbitrary term for small sawtimber trees. Generally, trees 12 to 18 inches in d.b.h. Also known as small sawtimber.

- **policies**
  statements on how the authority is to achieve its goals and objectives with regard to a specific subject area or class of subject areas, e.g., a policy for development on floodplains.

- **pollination**
  the movement of pollen from a stamen to a pistil by pollinators (birds or insects).

- **pollutant**
  a harmful chemical or waste material discharged into the water or atmosphere.

- **polygon**
  a closed geometric entity used to graphically represent area features with associated attributes.

- **Polymerase Chain Reaction (PCR)**
  a method that simulates an environment for rapid DNA replication of specific DNA segments. This procedure results in multiple copies of the specific DNA sequence which are then used in various experiments for diagnostic and analytical purposes.

- **polypeptide**
  1. chain of amino acids, usually less than 100 amino acids in length. Polypeptides are the building blocks of proteins.
  2. group of organisms that does not include their most recent common ancestor. (cf. monophyletic; technically speaking, the distinction between polyphyletic and paraphyletic is not as clear as it might seem.).

- **polytomous tree**
  a tree that has at least one branch point that is a polytomy. A tree with only one polytomy is called polytomous, as is a tree with multiple polytomies. This is in contrast to a dichotomous tree.

- **polytomy**
  a branch point on a tree that has more than two immediate descendents.

- **population**
  a group of individuals of one plant or animal taxon (species, subspecies or variety).

- **portable spar**
  a collapsible metal tower used in cable logging. AKA 'mobile spar or tower'.

- **post consumer**
  paper that has been recovered from the end consumer for recycling.

- **post-consumer waste**
  articles that have been used for their intended purpose and are ready to be discarded, such as boxes after they have been used to ship a product to its final destination. OCC is one type of post-consumer waste.

- **posterior**
  hind or hindmost; opposite of anterior.

- **potash**
  prepared by evaporating the lixivium of wood ashes in iron pots.

- **potential natural community**
  the biotic community that would be established if all successional sequences of its ecosystem were completed without additional man-caused disturbance under present environmental conditions. Grazing by native fauna, natural disturbances, such as drought, floods, wildfire, insects and disease are inherent in the development of potential natural communities which may include naturalised non native species.

- **potentially unstable soil area**
  any area where there is a moderate to very high likelihood of slope failure following conventional road construction or timber harvesting.

- **precipitation, average annual**
  the amount of rainfall (or equivalent snowfall) expected in the area over a calendar year.

- **precipitation, hourly**
  hourly amount of liquid equivalent precipitation.

- **precommercial thinning**
  1. see **juvenile spacing**.
  2. cutting trees from a young stand so that the remaining trees will have more room to grow to marketable size. Trees cut in a precommercial thinning have no commercial value and normally none of the felled trees are removed for utilisation. The primary in-

tent is to improve growth potential for the trees left after thinning.
3. when trees are about ten years old, teams go into the woods with saws and do a manual cleaning. This takes out the competing softwoods and thins the trees if they've grown up too densely.

- **pre-consumer waste**
leftover scrap generated in the paper and box manufacturing processes. Pre-consumer waste is clean waste that has been recycled by paper mills for decades.

- **predator**
an animal that preys on and devours other animals.

- **predator guard**
a physical barrier used to keep one animal from eating another. Usually refers to protection devices on nest boxes.

- **predisposition**
the weakening of an organism by some factor(s) of either the physical or biotic environment so as to render the organism more susceptible to a pathogen.

- **pre-harvest silviculture assessment (or survey)**
the survey carried out on a stand prior to logging to collect specific information on the silvicultural conditions such as planting survival, free-growing status, stocking, etc.

- **Pre-Harvest Silviculture Prescription (PHSP)**
a document that applies site-specific field data and develops forest management prescriptions for areas in advance of logging.

- **prehauling**
moving pulpwood from stump site to truck loading site by carrying it off the ground. Also known as forwarding.

- **preload**
to load logs or other wood products on to a trailer for subsequent transportation.

- **preprinted linerboard**
linerboard that is printed before being sent to a box plant for manufacture into a corrugated container. Boxes previously were only post-printed, or printed at the box plant after the wall of the container was

formed. Pre-printing allows great variety in design and colour choices.

- **prescribed burning**

1. the knowledgeable application of fire to a specific unit of land to meet predetermined resource management objectives.
2. controlled application of fire to wildland fuels in either their natural or modified state under specified environmental conditions which allow the fire to be confined to a predetermined area and at the same time to produce the intensity of heat and rate of spread required to attain planned resource management objectives.
3. skillful application of fire to natural fuels that allow confinement of the fire to a predetermined area and at the same time produce certain planned benefits.
4. a fire ignited under known conditions of fuel, weather, and topography to achieve specific objectives.

- **prescribed natural fire**

a fire ignited by natural processes (usually lightning) which is allowed to burn within specified parameters of fuels, weather and topography to achieve specific objectives.

- **prescribed or controlled burn**

the use of fire under specific environmental conditions to achieve forest management objectives. Used to reduce hazardous fuel levels, control unwanted vegetation, favour desired vegetation and improve visibility and wildlife habitat.

- **prescription**

a course of management action prescribed for a particular area after specific assessments and evaluations have been made.

- **present use valuation**

1. property tax relief classification based on the land's productivity for agriculture, horticulture, or forestry production, rather than for market value. Can result in substantial tax savings in areas where land values are high. Some restrictions and penalties apply, including a 3-year roll back provision with interest.

2. the action of reserving, protecting or safeguarding a portion of the natural environment from unnatural disturbance. It does not imply preserving an area in its present state, for natural events and natural ecological processes are expected to continue. Preservation is part of, and not opposed to, conservation.

3. an attempt to keep forests in an undisturbed state through the control of internal and external influences.

4. a management philosophy or goal which seeks to protect indigenous ecosystem structure, function, and integrity from human impacts. Management activities are generally excluded from 'preserved' forests.

- **preservative**

any substance applied to wood that is effective in preventing the development and action of wood destroying fungi and insects.

ACA: Ammoniacal Copper Arsenates.

ACC: Ammoniacal Copper Citrates.

ACQ: Ammoniacal Copper Quaternary.

ACZA: Ammoniacal Copper Zinc Arsenates.

CCA: Chromated Copper Arsenates.

PCP: Pentachlorophenol (Penta).

- **pressure-treating**

the process by which preservatives are applied to wood products. The treatment involves forcing the preservative, and sometimes water sealant, deep into the cellular structure of the wood under pressure in a closed cylinder.

- **prime mover**

heavy equipment used to tow other machines such as disc trenchers for site preparation.

- **principal defect**

the most significant defect that reduces tree volume and tree class.

- **prior appropriation**

the first-come first-serve approach to water use, most common in the western United States. The first user of water from a stream establishes a legal right for the continued use of the amount originally withdrawn. This use-it or lose-it approach offers no efficiency incentive.

- **prism**
  an optical instrument used as an angle gauge, consisting of a thin wedge of glass which establishes a fixed (critical) angle of projection in a point sample.

- **private ownership**
  forest land belonging to forestry companies or belonging to individual people or families.

- **probe**
  a short DNA fragment used to locate a specific sequence of DNA from a chromosome or genome.

- **problem forest type**
  non-merchantable forest types, including: stands of unfavourable stocking (i.e., dense small trees), low productivity sites and decadent stands with high waste and breakage.

- **procedure**
  a particular way of accomplishing an objective; generally refers to the method rather than the result. Procedures are usually developed to describe the methods for implementing policy.

- **proclamation date**
  the date on which a statute has legal effect.

- **producers**
  organisms that use solar (green plants) or chemical energy (some bacteria) to sustain life.

- **production forest**
  the forest used for production of various commodities, for example timber.

- **productive forest land**
  forest land that is capable of producing a merchantable stand within a defined period of time.

- **productivity**
  the ability of a stream, estuary, or the ocean to produce the food needed by anadromous fish at the various stages of their life cycle.

- **profile**
  a graphic representation of the rise and fall of a section of ground surface as seen in a vertical section view. Most commonly used in road de-

sign and cable payload analysis.

- **prokaryote**
unicellular organisms lacking a membrane-bound, structurally discrete nucleus and membrane-bound organelles. The term is generally applied to bacteria and archaea. Not a monophyletic group.

- **proleg**
any process or appendage that serves the purpose of a leg; specifically, the fleshy, unjointed, ventral abdominal projections of caterpillars and certain sawfly larvae.

- **protected area**
1. an area protected by legislation, regulation, or land-use policy to control the level of human occupancy or activities. Categories of protected areas include protected landscapes, national parks, multiple-use management areas and nature (wildlife) reserves.
2. areas such as provincial parks, federal parks, wilderness areas, ecological reserves, and recreation areas that have protected designations according to federal and provincial statutes. Protected areas are land and freshwater or marine areas set aside to protect the province's diverse natural and cultural heritage.

- **protection forest**
forest maintained on steep, unstable slopes to prevent accelerated erosion.

- **protein**
a large molecule composed of amino acid chains. Proteins perform a wide variety of cellular functions. Examples are hormones, enzymes and antibodies.

- **protist**
a paraphyletic group consisting of those eukaryotes which are not animals, true fungi or green plants.

- **protocol agreements**
an agreement between two or more ministries or two or more areas of the same ministry stating the role of each party in relation to the other or others with respect to an issue, or issues over which the parties have concurrent jurisdiction.

- **protozoan**
  from a phylogenetic point of view, an outmoded concept because it included unrelated eukaryotes - various amoebae, flagellates, ciliates and sporozoa.

- **provascular tissue**
  bundles of cells in young tree shoots.

- **provenance**
  the geographical area and environment to which the parent trees and other vegetation are native, and within which their genetic constitution has been developed through natural selection.

- **provincial forest inventory**
  a description of the quantity and quality of forest trees, non- wood values, and many of the characteristics of the land base compiled from statistical data for the forest lands of the province.

- **proximal**
  near to.

- **pruning**
  the manual removal, close to or flush with the stem, of side branches, live or dead and of multiple leaders from standing, generally plantation-grown trees. Pruning is carried out to improve the market value of the final wood product by producing knot-free wood for the improvement of the tree or its timber.

- **pseudopodium**
  transient extensions of the cell surface, used for locomotion or feeding. They may be supported internally (actinopods) or not (rhizopoda), they may be thread like (filose) or broad (lobose), may or may not bear extrusomes (nudipodia, extrusopodia) and they may be one (monopodial) to many (polypodial) produced at one time.

- **public**
  the entire population of British Columbia, including all organisations, companies, and groups.

- **public hearing**
  a hearing formally advertised and convened to afford any person who deems their interest in property to be affected by a proposal an opportunity to be heard by the Forest Service. The Forest

Service is not required to follow the tenor of the statements made at the hearing. A public hearing must be convened in respect of tree farm licence applications.

- **public highway**
  a highway for which public money has been spent and which is dedicated to public use by a plan deposited in the Land Titles Office for the district in which the road is situated.

- **public involvement**
  the procedures for obtaining and considering the views of the general public in planning and decision-making processes.

- **public trust doctrine**
  asserts that governments hold certain rights in trust for the public benefit and can take action to protect these rights from private interest.

- **pulley**
  see **sheave**.

- **pulp mill**
  a mill that manufactures pulp.

- **pulp timber**
  timber unsuitable for processing into lumber, because of its small diameter, or its condition.

- **pulpwood**
  1. wood used in the manufacture of paper, fiberboard, or other wood fiber products. Pulpwood sized trees are usually a minimum of 6 inches diameter at breast height.
  2. small timber bucked to length (can be longwood or shortwood lengths) for use as a fiber source in a pulpmill.

- **punctate**
  with a dimpled or spotted appearance.

- **pupa**
  the resting, inactive, nonfeeding instar in all holometabolous insects; the stage intermediate between the larva and the adult.

- **pure stand**
  a stand where at least 80% of the trees are of a single species in the main canopy.

- **pusule**
  a system involving a sac and channels and found in some dinoflagellates. The function is not understood, but it may act as an osmoregulatory organelle.

■ **pycnial stage (pycnium)**
a flaskshaped spore stage of the rust fungi; oozes out spores in a sticky matrix.

■ **pycnidium**
an open-pored, flaskshaped fruiting structure in which asexual spores called conidia are produced.

■ **pyrenoid**
a protein body lying inside some types of chloroplasts.

■ **quarry**
anything being hunted or pursued. A place where stone is excavated.

■ **quarter**
to fell trees across the hill at an angle rather than straight up or down the hill.

■ **quarter saw**
to saw a log into quarters lengthwise and then into boards in order to show the wood grain to advantage.

■ **quasi-judicial**
a decision made by a government official or tribunal which involves the application of policy to a particular set of facts requiring the exercise of discretion and the application of the principles of natural justice.

■ **quetzal**
a crested trogon bird of Central America.

■ **quorum**
the minimum number of members required to be present at an assembly or meeting before it can validly proceed to transact business.

■ **quotient**
the results obtained when one number is divided by another.

■ **radial growth (increment)**
the increase in tree radius over a period of time (e.g., 10 years or period between measurements) at breast height or occasionally at the base.

■ **radio controlled carriage**
a carriage that is operated by remote control with radio signals from the ground.

■ **radio whistles**
See '**talkie tooter**'.

■ **rain forest**
1. a tropical woodland that has an annual rainfall of at least 100 inches (254 centimetres) and often much more, typically restricted to certain lowland areas.

2. a dense, evergreen forest occupying a region having abundant rainfall throughout the year (often a tropical region).

■ **rain-on-snow events**
rainstorms that result in large amounts of surface runoff due to the combined effects of heavy rainfall and snow melt. Rapid snow melt is caused by heat supplied from the warm air that is characteristic of intense rainstorms and by heat released during condensation of moisture from the air onto the snow surface.

■ **raising lines**
the cables used to up-end a portable spar.

■ **rake**
a site preparation tool normally mounted on the front of a crawler tractor, used to remove trees, stumps roots and slash from a future planting site.

■ **range**
any land supporting vegetation suitable for wildlife or domestic livestock grazing, including grasslands, woodlands, shrub lands and forest lands.

■ **range enhancement**
any treatment, development, or structure designed to achieve or maintain the desired plant community.

■ **range of variability**
the spectrum of conditions possible in ecosystem composition, structure and function considering both temporal and spatial factors.

■ **range readiness**
the stage of plant growth at which grazing may begin under a specific management plan without permanent damage to vegetation or soil.

■ **range treatment class**
standard treatment categories identified in inventories of range resources to assess forage availability and treatments needed to improve forage production and/or protect the soil.

■ **range type**
a defined area with specific physical characteristics, which differs from other areas in its ability to produce distinctive kinds and amounts of vegetation and in its response to management.

range use plan
an operational plan that describes the range and livestock management measures that will be implemented to ensure that range resources are protected and that the management objectives for other identified resource values are achieved.

rangeland (range)
land on which vegetation is predominantly grasses, forbs, or shrubs suitable for grazing or browsing. Rangeland is generally and Shrubland, but may include some Treeland and Barren land. Agricultural land is excluded. Also included are areas seeded to native or adapted introduced species that are managed like native vegetation.

rangelands
a broad category of land characterised by native plant communities that are often associated with grazing. Rangelands are managed by ecological rather than agronomic methods.

rapattack
a method of initial attack whereby fire-fighters are transported to the fire in a specially-equipped helicopter, from which they descend during a hover by means of rappelling down a rope fitted with a mechanical device to control the rate of descent. In this way fire crews can be transported to fires which would otherwise pose difficult access problems.

rapids
stream habitats which possess a gradient greater than 4 percent with swiftly flowing water (> 50 cm/sec) having considerable turbulence. The substrate of rapids is generally coarser than the substrate of low gradient riffles, and during low streamflow conditions large boulders typically protrude through the surface.

rappel crew
an initial attack crew which rappels from a helicopter to access fires.

raptor
a bird of prey such as an owl, hawk, osprey or eagle.

- **rare species**
species which exist only in one or a few restricted geographic areas or habitats or occur in low numbers over a relatively broad area.

- **rasping**
a type of feeding by insects which rub or grate the leaf surfaces with their mouthparts to obtain particles for consumption.

- **Rate Of Spread (ROS)**
the speed at which a fire extends its horizontal dimensions, expressed in terms of distance per unit area of time. Generally thought of in terms of a fire's forward movement or head fire rate of spread, but also applicable to backfire and flank fire rate of spread.

- **rate-of-cut**
the proportion of the watershed area allowed to be cut each year.

- **reach**
1. a metal member connecting a logging trailer to the truck tractor.
2. the distance spanned by a skyline.
3. a length of stream channel, (lake or inlet) exhibiting, on average, uniform hydraulic properties and morphology.

- **reach number**
number given to a location on or segment of a stream within the Hydrologic Unit Code.

- **reaction wood**
wood with distinctive anatomical and physical characteristics, formed typically in parts of leaning or crooked stems and in branches, that tends to restore the original position of the branch or stem if this has been disturbed. Also known as tension wood (in broadleaved trees) and compression wood (in conifers).

- **recombinant DNA**
DNA constructed by joining DNA segments from two or more organisms. See **genetic engineering**.

- **reconnaissance**
the field examination of a proposed road location to determine its feasibility and possible impact on other resources, and to lay out the proposed centreline.

- **recovered paper**
  paper, used for its intended purpose, that is recovered from the waste stream for recycling.

- **recovery**
  that volume of timber which is recovered or saved after the felling and bucking operations are complete.

- **recreation feature**
  a biological, physical, cultural or historic feature that has recreational significance or value.

- **recreation feature objective**
  a resource management objective which reflects how a recreational feature or features will be managed, protected or conserved.

- **recreation feature significance**
  the quality, uniqueness, and availability of a recreation feature as classified in the recreation inventory.

- **recreation features inventory**
  one component of the Recreation Inventory. The identification, classification, and recording of the types and locations of biophysical recreation and cultural features, existing and potential recreation activities, feature significance and feature sensitivity.

- **recreation inventory**
  the identification, classification and recording of recreation features, visual landscapes, Recreation Opportunity Spectrum (ROS), recreation features of rivers and specific point locations of recreation sites, trails, caves etc.

- **recreation opportunity class**
  an assessment of the general potential of the site for outdoor recreation. The following minimum number of classes are recognised:
  Primitive: area characterised by essentially unmodified natural environment with a high probability of experiencing isolation from the sights and sounds of man.
  Roaded Natural: Area is characterised by a predominantly natural or natural-appearing environment with a low probability of experiencing isolation from the sights and sounds of man.
  Rural: Area characterised by a substantially modified

natural environment with a low probability of experiencing isolation from the sights and sounds of man.

Semi-primitive: Area characterised by a predominantly natural or natural-appearing environment with a moderate probability of experiencing isolation from the sights and sounds of man.

Urban: Area characterised by a substantially urbanised environment, although the background may have natural-appearing elements.

- **Recreation Opportunity Spectrum (ROS)**
  a mix of outdoor settings based on remoteness, area size, and evidence of humans, which allows for a variety of recreation activities and experiences. The descriptions used to classify the settings are on a continuum and are described as: rural, roaded resource, semi-primitive motorised, semi-primitive non-motorised, and primitive.

- **recreation opportunity spectrum objectives**
  resource management objectives in approved integrated resource management plans, reflecting the desired Recreation Opportunity Spectrum setting to provide for specific types of recreation opportunities and experiences.

- **recreation resource**
  a recreation feature, a scenic or wilderness feature or setting that has recreational significance or value or a recreation facility.

- **recreation use**
  the primary type of recreation use observed in the vicinity of the sample unit.

- **recreation value**
  see **recreation resource**.

- **recreation**
  any physical or psychological revitalisation through the voluntary pursuit of leisure time. Forest recreation includes the use and enjoyment of a forest or wildland setting, including heritage landmarks, developed facilities and other biophysical features.

- **recyclable paper**
  paper capable of being recycled. Almost all paper is recyclable if sorted by paper

type and if contaminants (such as plastic and food products) are removed.

- **recycled paper**
Papers made by manufacturers may contain various percentages of pre- and/or post- consumer waste, which may include materials (such as sawdust) that are viewed as solid waste.

- **redd**
a series of small nests for laying eggs built in bottom gravel by the female salmon.

- **red-listed species**
see **threatened or endangered species**.

- **reeve**
to thread a line through a block or other opening.

- **referral**
the process by which applications for permits, licences, leases, etc., made to one government agency by an individual or industry are given to another agency for review and comment.

- **reforestation**
1. establishing a forest by planting or seeding an area from which forest vegetation has been removed.
2. the natural or artificial restocking of an area with forest trees.
3. a process by which a forest is regrown.
4. the re-establishment of forest cover by natural or artificial means on areas recently supporting forest cover.
5. the re-establishment of trees on denuded forest land by natural or artificial means, such as planting and seeding.

- **regeneration**
1. the renewal of a tree crop through either natural means (seeded on-site from adjacent stands or deposited by wind, birds, or animals) or artificial means (by planting seedlings or direct seeding).
2. the renewal of vegetation by natural or artificial means.
3. the renewal of a crop whether by natural or artificial means.
4. the replacement of one forest stand by another as a result of natural seeding, sprouting, planting, or other methods; also young trees which will develop into the future forest.

- **regeneration cut**
  1. a cutting strategy in which old trees are removed while favourable environmental conditions are maintained for the establishment of a new stand of seedlings.
  2. a timber harvest designed to promote and enhance natural establishment of trees. Even-aged stands are perpetuated by three types of regeneration cuts: seed tree, shelter wood, and clearcutting. Uneven-aged stands are perpetuated by selecting individual or small groups of trees for removal (e.g., the selection system).

- **regeneration delay**
  the maximum time allowed in a prescription, between the start of harvesting in the area to which the prescription applies, and the earliest date by which the prescription requires a minimum number of acceptable well-spaced trees per hectare to be growing in that area.

- **Regeneration Performance Assessment (RPA)**
  a sampling survey carried out to collect field data on the height growth, competition, and stocking of young stands (5-10 years).

- **regeneration survey**
  carried out to determine the initial restocking of a site. It is used to describe the number of trees on a site that have reached acceptable standards.

- **regenerative brake**
  a device used on an interlock yarder which induces a retarding force on an outgoing line drum and transfers a portion of the power absorbed to an incoming line drum.

- **region/station/area**
  geographic grouping of forest service administrative units.

- **regional plan**
  the second level of planning in the Ministry of Forests hierarchical planning system. The regional forestry plan contains forest management alternatives based on a detailed analysis of timber supply within the region. Regional priorities for integrated use are identified and taken into account in setting production goals for

timber, range and forest recreation.

- **regionally important species**
species which are not red- or blue-listed, that require management practices that differ from standard integrated resource management guidelines in order to fulfill critical habitat needs; or locally or regionally threatened or declining species or those that may reasonably be expected to decline without protection of critical habitats.

- **registered lands**
a permit-only hunting program in which land is registered with and patrolled by the Wildlife Resources Commission. Hunters without a permit issued by the landowner are cited for trespass and prosecuted without need for the landowner to appear in court or swear out a warrant.

- **regular lay**
the most common type of wire rope used in logging. The wires making up the strands are laid in the opposite direction than the strands themselves.

- **regulation**
a rule, ordinance, or law by which conduct, etc. is regulated.

- **reinventory**
the complete restratification of an area on recent, mid-scale aerial photographs based on extensive field work.

- **relative humidity**
a ratio, expressed in percent, of the amount of moisture in a volume of air to the total amount which that volume can hold at the given temperature and atmospheric pressure.

- **release**
1. freeing a tree or group of trees from more immediate competition by cutting or otherwise eliminating growth that is overtopping or closely surrounding them.
2. removal of overtopping trees to allow understory or overtopped trees to grow in response to increased light.

- **reload**
to transfer logs from one mode of transportation to another or between vehicles.

- **relog**
  to salvage small timber, culls, or other residual wood products following the main logging operation.
- **Remote Automatic Weather Station (RAWS)**
  a weather station at which the services of an observer are not required. A RAWS unit measures selected weather elements automatically and is equipped with telemetry apparatus for transmitting the electronically recorded data via radio, satellite or by a landline communication system at predetermined times on a user-requested basis.
- **remote sensing**
  any data or information acquisition technique that utilises airborne techniques and/or equipment to determine the characteristics of an area.
- **renewable resource**
  a resource that has the capacity to be replaced through natural processes. Trees are a renewable resource.
- **reportable erosion event**
  a natural or man-made disturbance to the forest land base which is causing or will likely cause substantial environmental impacts, or which is a threat to life or property.
- **reprod**
  short for reproduction or young trees.
- **reproduction**
  1. the process by which young trees grow to become the older trees of the future forest.
  2. the process of forest replacement or renewal through natural sprouting or seeding or by the planting of seedlings or direct seeding.
  3. the generation of offspring by trees.
- **reserve**
  an area of forest land that, by law or policy, is not available for harvesting. Areas of land and water set aside for ecosystem protection, outdoor and tourism values, preservation of rare species, gene pool, wildlife protection etc.
- **reserve zone**
  the inner portion of a riparian management area situated adjacent to a stream,

lake, or wetland and established to conserve and maintain the productivity of aquatic and riparian ecosystems when harvesting is not permitted.

- **reserved trees**
  trees specifically reserved from harvesting and often referenced to cutting authorities or by map notations.

- **reserves**
  the retention of live or standing dead trees, pole size or larger, on site following harvest for purposes other than regeneration. Reserves can be uniformly distributed as single trees or left in small groups, and they can be used with any silvicultural system.

- **residual basal area**
  the basal area per hectare of acceptable trees left standing after harvest.

- **residual fibre**
  leftover wood fibre from sawmills and plywood plants, and pulpwood derived from timber management activities. By using residual fibres to produce paper, particleboard and medium density fibreboard, Willamette is able to fully utilise its resources.

- **residual stand**
  trees left in a stand to grow until the next harvest. This term can refer to crop trees or cull trees.

- **residual stand structure**
  the age class or height structure of the stand or remaining trees after harvesting.

- **residual tree**
  any tree not harvested within a harvest area.

- **residual trees (residual stand)**
  live trees left standing after the completion of harvesting.

- **residual value**
  the actual or assumed value of a machine after it has been fully depreciated. AKA 'salvage value'.

- **residue**
  the volume of timber left on the harvested area which meets or exceeds the size requirements but is below the log grade requirements of the minimum utilisation standards in the cutting authority. It is part of the allowable annual cut for cut control.

- **resilience**
  the ability of an ecosystem to maintain diversity, integrity and ecological processes following disturbance.

- **resistance to control**
  the relative ease of establishing and holding a fireguard and/or securing a control line as determined by the difficulty of control and resistance to fireguard construction.

- **resource**
  something that a country, state, etc., has and can use to its advantage (e.g. natural resources)

- **resource folio**
  a collection of resource capability and forest inventory maps, other resource data, interpretations, and management objectives for each resource sector. General prescriptions are developed to achieve the stated integrated use of objectives. A resource folio forms the basis for the timber licensee's development plan or working plan.

- **resource industry**
  an industry based on the primary resources obtained from agriculture, fisheries, forestry or mining.

- **resource integration**
  moving wood-based products to customers in several internalised, value-added steps.

- **Resource Management Zone (RMZ)**
  m regional or sub-regional plan: A division or zone of the planning area that is distinct from other zones with respect to biophysical characteristics, resource issues or resource management direction. Resource management zones may be drawn on a map to describe general management intent. The zones are usually further defined using descriptive objectives and strategies to explain future land use and resource management activities.

- **resource value**
  values on crown land which include but are not limited to biological diversity, fisheries, wildlife, minerals, oil and gas, energy, water quality and quantity, recreation and tourism, natural and cultural heritage, timber,

forage, wilderness and aesthetic values.

- **responsibility**
condition, quality, fact, or instance of being responsible; obligation, accountability, dependability, etc.

- **restocking**
replanting trees after felling.

- **restoration**
the return of an ecosystem or habitat to its original community structure, natural complement of species and natural functions.

- **retention**
retaining or saving a portion of the original stand in a cluster or clump. Retention visual quality objective: a visual landscape strategy derived from landscape analysis which applies to areas of high landscape value (for example, continuously forested or steep slopes facing important viewpoints or recreation use areas, foreground areas adjacent to important viewpoints or recreation use areas, and certain shorelines). Forest management activities may be present, but should not be noticed by the average viewer. Some visual change may be discernible, but should not be recognised as being different from existing natural features in the landscape.

- **rhizomorph**
a thread- or cordlike fungal structure made up of hyphae.

- **riffle**
1. a shallow rapids where the water flows swiftly over completely or partially submerged obstructions to produce surface agitation, but standing waves are absent. Three types of riffle habitats are identified. Low gradient riffles are shallow (< 20 cm deep) stream reaches with moderate current velocity (20-50 cm/sec) and moderate turbulence. Substrate is usually composed of gravel, pebble, and cobble-sized particles. The upper gradient for this habitat is arbitrarily limited to 4 percent.
2. a shallow stream area characterised by broken water surface in coarser substrates.

- **rigging**
the cables, blocks and other equipment used in yarding and loading logs.

- **rigging slinger**
  the headman on a rigging crew who is responsible for choker setters and chaser and who selects the logs to make up a turn.

- **right-of-way**
  the strip of land over which a power line, railway line, road, etc., extends.

- **rig-up**
  to prepare a tailhold, spar, or tower for yarding by guying and anchoring it, attaching all rigging and stringing the lines.

- **rig-up block**
  a small block through which the straw line is strung in the brush; usually wide throated so that splices and connectors for the straw line segments can pass through the block.

- **rim pull**
  the torque that a machine can exert at the contact of its drive wheels to the ground; that is the tractive effort that a machine can exert to move itself.

- **ringshake**
  peripheral cracks in woody tissues of stems. The pattern of damage is concentric with the annual rings.

- **ringspot**
  a circular area of chlorosis with a green centre.

- **riparian**
  1. the land adjacent to and pertaining to the banks of streams, rivers, or other water bodies.
  2. an area of land adjacent to a stream, river, lake or wetland that contains vegetation that, due to the presence of water, is distinctly different from the vegetation of adjacent upland areas.

- **riparian habitat**
  vegetation growing close to a watercourse, lake, swamp, or spring that is generally critical for wildlife cover, fish food organisms, stream nutrients and large organic debris, and for streambank stability.

- **riparian rights**
  a system of water rights that gives anyone whose land adjoins a flowing stream the right to use water from the stream as long as some is left for downstream landowners.

- **riparian zone**

an area adjoining a body of water, normally having soils and vegetation characteristic of floodplains or areas transitional to upland zones. These areas help protect the water by removing or buffering the effects of excessive nutrients, sediments, organic matter, pesticides or pollutants.

- **riprap**

1. an apron of coarse rock installed over the fill slope to prevent erosion.
2. a layer of boulders or shot rock fragments placed over soil to protect it from the erosive forces of flowing water or waves.

- **risk**

the probability of an undesirable event occurring within a specified period of time. With regard to insect populations, risk involves components to evaluate the likelihood of an outbreak, the likelihood of trees being attacked (susceptibility) or the likelihood of trees being damaged (vulnerability). In fire prevention, risk involves those things or events that cause fires to start (including the physical igniting agents and people).

- **risk assessment**

a quantitative and qualitative approach to monitoring the hazardous capacity of a new product. This involves the identification and characterisation of hazards, an assessment of exposure to the product and a final risk characterisation of the product.

- **risk rating**

the process of identifying the degree of risk that timber harvesting imposes on adjacent and downslope social, economic and forest resource values. The severity of each potential hazard and the magnitude of the potential consequences that correspond to each hazard provide the overall risk associated with harvesting a site.

- **ribonucleic acid (RNA)**

a molecule found in the nucleus and cytoplasm of cells. It plays an important role in protein synthesis and other chemical processes of the cell. The structure of RNA is similar to that of DNA. There are several classes of RNA molecules, including messenger RNA, transfer RNA, ribosomal RNA, and other small RNAs, each serving a different purpose.

- **road**
  the path followed by a turn of logs yarded by a cable system. AKA 'yarding road', 'skyline road', or 'skyline corridor'.

- **road change**
  see **line change**.

- **road deactivation**
  measures taken to stabilise roads and logging trails during periods of inactivity, including the control of drainage, the removal of sidecast where necessary and the re-establishment of vegetation for permanent deactivation.

- **road functional class**
  alpha code designating the functional classification to road segment. The classification is as follows:
  Arterial: A road segment that provides service to large land areas and usually connect with other arterial roads or public highway.
  Collector: Segments that connect major heavily travelled multiple purpose arterial routes and single source local roads.
  Local: Segments which connect terminal facilities such as log landings and recreation sites with Forest collector or arterial roads or with public highways. Usually single purpose.

- **road location line**
  the marked location of proposed roads.

- **road prism**
  the area of the ground containing the road surface, cut slope and fill slope.

- **road surface**
  an alpha code describing the general type of running surface on an existing road management section.

- **road system**
  an alpha code indicating primary systems designation where primary indicates the system under which principle funding and management criteria for operation and maintenance of a road is derived.

- **roads**
in logging, generally classified as skid trails, spur roads and main haul roads.

- **roe**
salmon eggs.

- **rolling resistance**
the retarding force exerted by the ground against a vehicle's wheels.

- **root**
the part of a plant, usually below the ground, that lacks nodes, shoots, and leaves; holds the plant in position; draws water and nourishment from the soil and stores food.

- **root collar**
the transition zone between stem and root at the ground line of a tree or seedling.

- **root hairs**
threadlike extensions that grow from a plant root and takes in water and minerals from the soil.

- **root rot**
disease that destroys tree roots, often killing the tree.

- **rootwad**
the mass of roots associated with a tree adjacent or in a stream that provides refuge and nutrients for fish and other aquatic life.

- **ROPS**
Roll Over Protection Structure. A sturdy structure built into or fitted around the operator cab on certain equipment which protect operators if the machines overturn.

- **rotation**
1. the planned number of years between the formation or regeneration of a tree crop or stand and its final cutting at a specified stage of maturity. Can be based on physical, biological, pathological or economic criteria.
2. the number of years required to establish and grow trees to a specified size, product, or condition of maturity. A pine rotation may range from as short as 20 years for pulpwood to more than 60 years for sawtimber.
3. the time required to establish and grow trees to a certain product size. A pine rotation may be as short as 20 years for pulpwood, and

60 years or more for sawtimber.
4. the planned time interval between regeneration cuts in a forest.

■ **rotation age**
the age at which a stand is considered mature and ready for harvesting.

■ **rough cut**
lumber that has not been dressed (surfaced) but which as been sawn, edged, and trimmed to at least show saw marks in the wood on the four longitudinal surfaces of each piece for its overall length.

■ **roundwood**
1. sections of tree stems, with or without bark. Includes logs, bolts, posts and pilings.
2. a length of cut tree generally having a round cross-section, such as a log or bolt.

■ **roundwood products**
logs, bolts or other round sections cut from trees for industrial or consumer use.

■ **rub tree**
a tree used as a fender or pivot for a moving line to protect the remaining stand during yarding. Usually will be so badly damaged after use that it must be cut down.

■ **rules**
informal working term for draft forest practices requirements proposed for the forest practices.

■ **running guy**
see **sail guy**.

■ **running line**
a moving cable.

■ **running skyline**
a yarding system with three suspended moving lines, generally referred to as the main, haulback, and slack pulling, that when properly tensioned will provide lift and travel to the carriage.

■ **rut or rutting**
depressions in the soil caused by repeated machine traffic.

■ **safe working load**
the load a wire rope or structure may safely carry. Usually the ultimate or breaking load divided by the desired factor of safety.

■ **sag**
see **deflection**.

- **sail guy**
  a guy line attached to end of the leaning tower on a swing yarder. The guy goes slack when the boom swings. AKA 'running guy'. Largely made obsolete on modern swing yarders by the gantry tower which accomodates the guylines for the yarder.

- **sale unit**
  a timber sales arrangement in which the buyer pays for forest products removed in units (measured in cords, MBF, or units of weight). Determination of units removed from the area is verified by mill tally, scale tickets and buyer's or seller's tally.

- **sale, lump sum (boundary)**
  the sale of specified timber on a specified area. The volume may or may not be estimated and published. The buyer is responsible for determining correct volume. The seller guarantees ownership and boundaries.

- **sale, lump sum (boundary)**
  the sale of specified timber on an a specified area. The volume may or may not be estimated and published. The buyer is responsible for determining correct volume. The seller guarantees ownership and boundaries.

- **salmonid**
  a fish of the fish family Salmonides; for example salmon, trout and chars.

- **salvage**
  the harvesting of dead and dying timber while it still has commercial value is called 'salvage.' Trees impacted by fire, insects and disease can often be salvaged.

- **salvage cut**
  1. removal of dead, damaged, or the harvesting of dead or damaged trees or of trees in danger of being killed by insects, disease, flooding, or other factors in order to save their economic value.
  2. the diseased trees with the intent of recovering value prior to deterioration.

- **salvage harvesting**
  logging operations specifically designed to remove damaged timber (dead or in poor condition) and yield a wood product. Often carried out following fire, insect attack or windthrow.

■ **salvage value**
see **residual value**.

■ **sand**
loose soil made up of small rock particles.

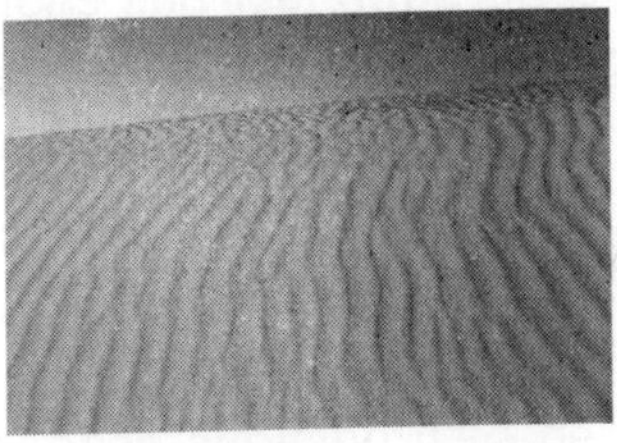

■ **sanitation treatment**
tree removal or modification operations designed to reduce damage caused by forest pests and to prevent their spread.

■ **sapling**
1. a loose term for a young tree no longer a seedling but not yet a pole, about 1 - 2 m high and 2 - 4 cm DBH, typically growing vigorously and without dead bark or more than an occasional dead branch. Also, a young tree having a DBH greater than 1 cm but less than the smallest merchantable diameter.
2. a small tree, usually between 2 and 4 inches diameter at breast height.
3. a young tree.

■ **saprobe**
an organism that secures its food directly from nonliving organic matter.

■ **saprophyte**
an organism living on dead organic matter.

■ **sapwood**
the light-coloured wood that appears on the outer portion of a cross-section of a tree.

■ **sapwood**
1. the younger wood nearer to the surface of the tree, which does not have the natural extractives (chemicals) to prevent decay.
2. the soft wood between the inner bark of a tree and the heartwood, serving to conduct water.
3. the outer part of the wood; also called alburnum.

■ **saw log**
1. a log suitable in size and grade for producing sawn lumber.
2. a log large enough to yield lumber. Usually the small end of a sawlog must be at least 6 to 8 inches in diameter for softwoods and 10 to 12 inches for hardwoods.

- **sawlog length**
the length in feet of the merchantable portion of a sawtimber growing-stock tree.

- **sawmill**
a mill that cuts logs into lumber of different sizes

- **scale**
an instrument which measures the weight or volume of a log or load of logs.

- **scale deduction**
the volume deducted for rot, breaks, or insufficient trim, from a log's gross scale.

- **scaling**
1. the measuring of lengths and diameters of logs and calculating deductions for defect to determine volume.
2. determination of the gross and net volume of logs using the customary commercial volumetric units for the product.
3. involved.

- **scalping**
1. site preparation method which exposes favourable mineral soil for tree seedlings to be planted in.
2. removing small plants and duff or ashes from around the spot where a tree seedling will be planted. Usually done by hand rather than by machine.

- **scarification**
1. a method of seedbed preparation which consists of exposing patches of mineral soil through mechanical action.
2. shallow loosening of the soil surface.
3. the process of disrupting or removing the woodland floor, or mixing it with the mineral soil by mechanical action preparatory to natural or direct seeding or the planting of tree seedlings.

- **scarifier**
a device used to churn up soil before seedlings are planted on a site.

- **scarify**
weakening of a seed shell, enabling it to germinate.

- **scarifying**
1. for soil: The removal of the top litter layer of an area (usually in strips) for site preparation.
2. for seed: The abrasion or weakening of the seed coat to encourage germination.
3. churning up the soil before seedlings are planted.

■ **scenic area**
any visually sensitive area or scenic landscape identified through a visual landscape inventory or planning process carried out or approved by the district manager.

■ **school marm**
a forked top tree with two tops.

■ **scion**
an unrooted portion cut from a plant to make a graft.

■ **scorch**
appearing as if tissues were burned by heat; usually affecting marginal portions of leaves.

■ **screefing**
removal of herbaceous vegetation and soil organic matter to expose a soil surface for planting.

■ **sealant**
a water repellent which may be forced into the wood along with the chemical preservative in a closed cylinder under pressure. However, treated wood should be cleaned and resealed yearly to maintain optimum appearance.

■ **sealed bid sale**
sale in which interested parties submit written bids at the time and place specified.

■ **seasoned**
wood that has been dried to a certain moisture content to improve its serviceability.

■ **seasoning**
wood is both air-dried and steam kiln-dried to reduce the water content.

■ **second growth**
1. a forest or stand that has grown up naturally after removal of a previous stand by fire, harvesting, insect attack or other cause.
2. trees that come up naturally after the first growth of timber has been cut or destroyed by fire. Also known as young timber.
3. the forests re-established following the removal of virgin (i.e., previously unharvested) or old-growth stands. Most of Massachusetts forests are either second or third growth.

■ **second pass**
the next entry to harvest timber after green-up (or other recovery objective) occurs.

▪ **secondary channel**
subordinate channel in a stream reach with more than one channel; minor channel in a floodplain.

▪ **secondary logging road**
road designed for relatively little use. Typically a dirt road, with no gravel, used only during dry weather.

▪ **section**
land survey subdivision. Usually one square mile (640 acres).

▪ **sediment**
1. suspended or deposited soil and organic matter in streams that originates from erosion processes.
2. soil material that has been eroded from the land surface, often by overland water flow and then transported and deposited away from its original location.

▪ **sedimentation**
1. the deposition of eroded soil material (sediment) into an ephemeral, intermittent or perennial stream, lake, or other water body.
2. the process of sediment deposition usually resulting from erosion.
3. the process of subsidence and deposition by gravity of suspended matter carried in water; usually the result of the reduction of water velocity below the point at which it can transport the material in suspended form.

▪ **seed**
the part of a flowering plant that typically contains the embryo with its protective coat and stored food and that will develop into a new plant, if sown.

▪ **seed coat**
a protective covering around the seed.

▪ **seed orchard**
1. a plantation of specially selected trees that is managed for the production of genetically improved seed.
2. orchards where trees are grown for the specific purpose of producing seeds.

▪ **seed source**
the locality where a seed lot was collected. If the stand from which collections were made was exotic, the place where its seed originated is the original seed source.

▪ **seed tree cut**

1. a harvesting method in which a few scattered trees are left in the area to provide seed for a new forest stand. Selection of seed trees should be based upon growth rate, form, seeding ability, wind firmness, and future marketability. This harvesting method produces an even-aged forest.
2. a regeneration cut where mature trees are left standing in a harvested area to provide seed for regeneration of the cutover site.

▪ **seed tree silvicultural system**

an even-aged silvicultural system in which selected trees (seed trees) are left standing after the initial harvest to provide a seed source for natural regeneration. Seed trees can be left uniformly distributed or in small groups. Although regeneration is generally secured naturally, it may be augmented by planting. Seed trees are often removed once regeneration is established or may be left as reserves.

▪ **seed trees**

trees selected to be left standing to provide seed sources for natural regeneration. Selection is usually on the basis of good form and vigour, the absence of serious damage by disease, evidence of the ability to produce seed and wind firmness.

▪ **seed year**

a year in which a given species produces a large seed crop over a sizable area. Some species of trees produce seeds irregularly.

▪ **seedbed**

1. in natural regeneration, the soil or forest floor on which seed falls; in nursery practice, a prepared area over which seed is sown.
2. area prepared to receive seeds, such as an area cleared of plants and duff.
3. new forest.

▪ **seedling**

1. a young tree, grown from seed, from the time of germination to the sapling stage, having a DBH equal or less than 1 cm.
2. a tree, usually less than 2 inches diameter at breast

height, that has grown from a seed rather than from a sprout.
3. a nursery-grown tree that has not been transplanted in the nursery.
4. the number of established tree seedlings or shrub stems on a fixed area or sample plot.

- **seedlot**
1. a quantity of cones or seeds having the same species, source, quality and year of collection.
2. seed from a particular collection event, either from a single tree collection or a pooling of seed from many trees.

- **seepage zone**
an area on a hill slope or at the slope base where water frequently or continuously springs to the surface.

- **seismic line**
a constructed trail used for seismographic exploration.

- **selection cut**
a regeneration cut designed, to create and perpetuate an uneven-aged forest. Trees may be removed singly or in small groups. A well-designed selection cut removes trees of lesser quality and trees in all diameter classes along with merchantable and mature high-quality sawlog trees. Should be differentiated from 'select' or 'selective' cuts, which often equate to high-grading.

- **selection cutting**
an uneven aged silviculture system in which trees are harvested individually or in small groups continuously at relatively short intervals.

- **selection method**
removal of the mature timber, usually the oldest or largest trees, either as single scattered individuals or in small groups, at relatively short intervals repeated indefinitely in order to encourage continuous reproduction and uneven-aged stands.

- **selection pressure**
environmental influences or physical characteristics of an organism that determine its likelihood of being selected among its co-habitants.

- **selection silvicultural system**
a silvicultural system that removes mature timber ei-

ther as single scattered individuals or in small groups at relatively short intervals, repeated indefinitely, where the continual establishment of regeneration is encouraged and an uneven-aged stand is maintained.

- **selection system**
  uneven-aged silvicultural system in which single or small groups of trees are periodically selected to be removed from a large area so that age and size classes of the reproduction are mixed.

- **selection thinning**
  removal of dominants that have exceeded the diameter limit prescribed, in favour of smaller trees with good growth form and condition. This will promote conversion to a selection forest.

- **selective cutting**
  the periodic removal of individual trees or groups of trees to improve or regenerate a stand. See '**selection method**'

- **selective logging**
  removal of certain trees in a stand as defined by specific criteria (species, diameter at breast height, or height and form). It is analogous to highgrading. Not to be confused with the selection silvicultural system.

- **self-loading logging truck**
  a logging truck with its own loading device, generally a knuckleboom mounted behind the cab. AKA 'a self loader'.

- **self-motivation**
  the urging or inciting of oneself to action.

- **semi-permanent bridge**
  a bridge having a substantial proportion of its components constructed of steel, concrete, or timber that has been pressure-treated with a suitable preservative.

- **senescence**
  1. aging of tissues; growing old.
  2. death of a leaf triggered by an increase in the enzymes that promote the breakdown of plant cells. Begins when shorter days and cooler temperatures occur.

- **sensitive areas**
  small areas designated to protect important values

during forest and range operations. These areas, established by a Ministry of Forests district manager in consultation with a designated environment official, guide operations on a site- specific basis and require a combination of forest practices. Sensitive areas will be mapped by resource agencies and include regionally significant recreational areas, scenic areas with high visual quality objectives and forest ecosystem networks.

▪ **sensitive areas objectives**
to adequately manage, protect and conserve the resources of the area. Sensitive areas may be designated through a planning process, or by the ministry of forests district manager and designated environment official (for example, forest ecosystem networks and the setting of visual quality objectives for sensitive scenic areas).

▪ **sensitive resource area**
an identifiable geographic unit of the forest land base that requires a specific combination of forest practices to adequately protect important resource values.

▪ **sensitive slopes**
any slope identified as prone to mass wasting.

▪ **sensitive soils**
forest land areas that have a moderate to very high hazard for soil compaction, erosion, displacement, mass wasting or forest floor displacement.

▪ **sensitive watershed**
a watershed that is used for domestic purposes or that has significant downstream fisheries values and in which the quality of the water resource is highly responsive to changes in the environment. Typically, such watersheds lack settlement ponds, are relatively small, are located on steep slopes, and have special concerns such as extreme risk of erosion.

▪ **sensitive/vulnerable species**
sensitive or vulnerable species as identified by the ministry of environment, lands and parks. Blue-listed species are considered to be vulnerable and 'at risk' but not yet endan-

gered or threatened. Populations of these species may not be declined by their habitat or other requirements are such that they are sensitive to further disturbance. The blue list also includes species that are generally suspected of being vulnerable, but for which information is too limited to allow designation in another category.

- **sepals**
  the separate parts of the flower calyx that is the outside covering of a bud.

- **sequence**
  the linear order of the nucleotides on the DNA strand. This sequence determines the unique DNA composition of an organism.

- **seral stage**
  1. any stage of development of an ecosystem from a disturbed, unvegetated state to a climax plant community.
  2. the series of changes occurring in the ecological succession of a plant community, e.g. pioneer stage or climax stage.

- **serotinous**
  a pine cone or other seed case that requires heat from a fire to open and release the seed.

- **sessile**
  1. attached or fastened; incapable of moving from place to place.
  2. refers to organisms which are fixed to the substrate, for example by means of a stalk or lorica.

- **set**
  a cutting crew, generally composed of two people, a faller and a bucker. AKA 'felling set'.

- **seta**
  slender, hair like or bristly projections arising from the epidermal layer on any part of the body of an insect.

- **setback**
  when a tree settles back on the backcut, opposite the direction in which it was supposed to fall.

- **setting**
  the area yarded or skidded to one landing.

- **settlement pond**
  larger than a catchment basin and preferably with lower

velocity water flows that enable suspended sediment to settle before the flow is discharged into a creek.

- **sexual stage**
  reproductive stage of the life cycle of an organism.

- **shackle**
  a U-shaped metal connector with a pin or threaded bolt through the ends. AKA 'clevis'.

- **shade tolerance**
  the capacity of a tree or plant species to develop and grow in the shade of, and in competition with, other trees or plants.

- **shade-intolerant trees**
  trees that cannot thrive in the shade of larger trees.

- **shade-tolerant**
  the capacity of a tree or plant species to develop and grow in the shade of, and in competition with, other trees or plants.

- **shay swivle**
  a fitting used to attach the slack-pulling line to the mainline on a skyline system.

- **shear**
  a mechanical cutting head, attached to a harvester, which severs the tree from the stump with a scissors-like action. Usually an undesirable cutting head for saw timber as it often splits the butt log and degrades the value of that log.

- **shearing**
  1. in Christmas tree culture, to prune the branches to make dense foliage and give the tree a conical shape.
  2. slicing or cutting trees or stumps at the ground line. Shearing may be done at harvest or with a KG blade during site preparation.

- **sheave**
  a wheel with a groved rim. AKA 'pulley'.

- **shelter wood**
  1. the silvicultural method that gradually removes the trees from a given stand by a series of partial cuttings over a relatively short period of time. Eventually all overstory trees are removed, resulting in an even-aged stand.
  2. a method of harvesting in which some mature trees are left standing in order to provide protec-

tion and shelter for the young regeneration.
3. a regeneration cut designed to stimulate reproduction by removing all overstory trees. This is achieved by a series of cuts over several years. Gradual reduction of stand density protects understory trees and provides a seed source for stand regeneration.

- **shelter wood cut**
  removing trees on the harvest area in a series of two or more cuttings so new seedlings can grow from the seed of older trees. This method produces an even-aged forest.

- **shelterwood logging**
  method of harvesting timber so that selected trees remain scattered throughout the tract to provide seeds for regeneration and shelter for seedlings.

- **shelterwood silvicultural system**
  a silvicultural system in which trees are removed in a series of cuts designed to achieve a new even-aged stand under the shelter of remaining trees.

- **shelterwood system**
  even-aged silvicultural system in which a new stand is established under the protection of a partial canopy of trees. The mature stand is generally removed in a series of two or more cuts, the last of which is when the new even-aged stand is well developed.

- **shipbuilding**
  a business that designs and builds ships

- **shortwood**
  pulpwood less than 10 feet long.

- **shortwood system**
  a harvesting system in which the tree is felled, delimbed and cut into short lengths before it is hauled to the landing (usually by a forwarder).

- **shotgun**
  a two-drum, live skyline yarding system in which the carriage moves down the skyline by gravity, is lowered to attach logs, then raised and pulled to the landing by the main line. AKA 'flyer'.

■ **shotholes**
small holes in a leaf caused by feeding activity and giving the appearance of injury via a shotgun.

■ **shovel logging**
relaying logs from a harvest area to roadside using a hydraulic log loader.

■ **show**
any unit of operation in the woods, usually associated with logging, such as a highlead show.

■ **shrub**
a plant that has persistent woody stems and a relatively low growth habit, and that generally produces several basal shoots instead of a single bole. It differs from a tree by its low stature and non-arborescent form. Usually less than 16 ft. tall at maturity.

■ **shrubland**
areas on which vegetation is dominated by shrubs, provided these areas do not qualify as Built-up land or Cultivated cropland. Examples are areas consisting of such species as big sagebrush, shadscale, greasewood, creosotebush and fourwing saltbrush.

■ **side**
1. a group of people and machines working together as a unit. A contractor might have a 'skidder side' and a 'cable side' working on the same 'show.'
2. a logging unit-the workers and equipment needed to yard and load any one unit of operation. e.g. 'The company had a cable and skidder side operating in the woods.'

■ **side slope**
the average slope of the ground, in percent, along a yarding road on either side of a centreline.

■ **sideblocking**
a method of laterally displacing a slack line to facilitate log hook up or permit yarding a wider road. AKA 'Dutchman block'

■ **sidecast**
moving excavated material onto the downslope side of a temporary access structure, excavated or bladed trail, or landing during its construction.

- **side hill yarding**
  yarding logs parallel, or nearly so, to the ground contour lines.

- **sign**
  the actual presence of the causal organism in association with the disease symptoms.

- **siliceous**
  incorporating silica.

- **sills**
  a single structural member used as a foundation to transfer the loads from the bridge superstructure to the supporting soil.

- **silt fence**
  a temporary barrier used to intercept sediment-laden runoff from small areas.

- **silvics**
  the study of the life history, requirements and general characteristics of forest trees and stands in relation to the environment and the practice of silviculture.

- **silvicultural system**
  a planned program of treatments throughout the life of the stand to achieve stand structural objectives based on integrated resource management goals. A silvicultural system includes harvesting, regeneration and stand-tending methods or phases. It covers all activities for the entire length of a rotation or cutting cycle.

- **silvicultural system variant**
  further describes the functional attributes of a silvicultural system. A variant describes the general distribution of cut-and-leave areas or leave-trees on an area over time. The silvicultural system variant is the recommended minimum level of description of a silvicultural system for a silvicultural prescription or stand management prescription.

- **silviculture**
  1. the art and science of controlling the establishment, growth, composition, health and quality of forests and woodlands. Silviculture entails the manipulation of forest and woodland vegetation in stands and on landscapes to meet the diverse needs and values of landowners and society on a sustainable basis.

2. the science and practice of controlling the establishment, composition and growth of the vegetation of forest stands.
3. the art and science of growing and tending a forest (managing stands of trees to achieve desired outcomes relative to species composition and stand structure).
4. the theory and practice of controlling woodland establishment, composition, structure and growth. Silvicultural practices consist of the various treatments that may be applied to woodland stands to maintain and enhance their utility for any purpose.

- **silviculture prescription**

a site-specific operational plan that describes the forest management objectives for an area. It prescribes the method for harvesting the existing forest stand and a series of silviculture treatments that will be carried out to establish a free growing stand in a manner that accommodates other resource values as identified.

- **silviculture regime**

a series of site-specific silviculture treatments planned over time.

- **silviculture survey**

a sampling procedure to determine silvicultural conditions such as planting survival, free-growing status, stocking, etc., leading to management decisions.

- **silviculture treatment**

1. any silviculture activity on forest stands to meet stand-specific objectives.
2. activities that ensure the regeneration of young forests on harvested areas and enhance tree growth and improve wood quality in selected stands.

- **simple**

a single leaf blade with a bud at the base of the leaf stem.

- **single tree selection**

see **selection silvicultural system**.

- **singlejacking**
when one person does the felling and bucking on a setting.

- **single-span skyline**
a skyline without intermediate supports. Single span skylines are classified as short span if they reach out less than 2000 ft to the tailblock and long span if they exceed 2000 ft. Note, External yarding distances may be shorter than the spar to tailblock distance.

- **sister group**
the two groups resulting from the splitting of a single lineage.

- **site**
1. an area described or defined by its biotic, climatic, and soil conditions in relation to its capacity to produce vegetation; the smallest planning unit.
2. an area having some capacity to produce a particular type of timber species or other vegetation based on the combination of biological, climatic and soil and soil factors present. Site expresses the capacity of a given setting to grow timber or other vegetation.
3. the combination of biotic, climatic, topographic and soil conditions of an area; the environment at a location.

- **site class**
the measure of the relative productive capacity of a site for a particular crop or stand, generally based on tree height at a given age and expressed as either good, medium, poor or low.

- **site index**
1. an expression of the forest site quality of a stand, at a specified age, based either on the site height, or on the top height, which is a more objective measure.
2. a relative measure of forest site quality based on the height (in feet) of the dominant trees at a specific age (usually 25 or 50 years, depending on rotation length). Site index information helps estimate future returns and land productivity for timber and wildlife.
3. height of a tree at a specified index or base age. Used

as an indicator of site quality.

- **site preparation**
 1. any scheduled operation that is used to prepare an area for the re-generation of trees. These operations can be used individually or together, as needed, to prepare a site for tree regeneration. Examples include: piling unusable woody debris or slash burning excess leftover branches and limbs disking the ground bedding or furrowing applying forest chemicals.
 2. the treatment of the soil and ground vegetation to prepare the soil surface as a favourable seedbed for either naturally or artificially disseminated seed or for planted seedlings.

- **site productivity class**
 a classification of forest lands in terms of inherent capacity to grow crops of industrial wood. The class identifies the potential growth in merchantable cubic feet/acre/year and is based on the age of culmination of mean annual increment of fully stocked natural stands.

- **site quality**
 the inherent productive capacity of a specific location (site) in the forest affected by available growth factors (light, heat, water, nutrients, anchorage); often expressed as tree height at a given age.

- **site rehabilitation**
 the conversion of the existing unsatisfactory cover on highly productive forest sites to a cover of commercially valuable species.

- **site sensitivity**
 an assessment of the susceptibility of a site to soil-degrading processes, such as soil compaction, erosion, mass wasting and forest floor displacement.

- **site tree quality**
 a classification of sample tree according to how well the tree reflects the productive potential of the site.

- **site-specific**
 pertaining to a specific planning unit.

- **situation report (SITREP)**
 an itemised list and/or written account, usually issued on a

daily basis, detailing the status of various fire-related activities. A SITREP generally contains information on fire occurrence and area burned to date, fire suppression resources committed to going fires and resources on standby, number of fires in the various stages of control, fire danger class, fire weather forecast and forest closures (if any).

- **siwash**
  a line not running in a straight line by being bent around a tree, stump, rock, etc.

- **size down woody material**
  dead twigs, branches, stems and boles of trees and brush that have fallen and lie on or above the ground but within 1.8, (6 feet) meters of the ground. The various kinds of woody material are listed below:
  Small Twigs: Small twigs are defined as those fuel pieces which have a cross section diameter of less than 1/4 inch (6mm) at the point of intersection with the sampling plane.
  Large Twigs: Large twigs are defined as those fuel pieces which have a cross section diameter of between .25 and .99 inches (6 and 25mm) inclusive at the point of intersection with the sampling plane.
  Branches: Branches are defined as those fuel pieces with a cross section diameter of between 1.0 and 2.99 inches (25 and 75mm) inclusive at the point of intersection with the sampling plane.
  Large Fuel: Branches or pieces 3.0. (75mm) or larger at the point of intersection with the sampling place. The actual diameter is usually recorded for each large piece selected.

- **skeletoniser**
  insects which consume leaf tissue, often from the lower side of the leaf, leaving the upper epidermis and vascular tissues intact.

- **skid**
  the bunch of logs pulled behind a track or wheel skidder. Also known as a pull, turn, drag or twitch among other names.

- **skid road**
  a bladed or backhoe-constructed pathway where

stumps are removed within the running surface as necessary. Skid roads are suitable only for tracked or rubber-tired skidders bringing trees or logs from the felling site to a landing.

- **skid trail**

a random pathway travelled by ground skidding equipment while moving trees or logs to a landing. A skid trail differs from a skid road in that stumps are cut very low and the ground surface is mainly untouched by the blades of earth moving machines.

- **skidder**

1. a wheeled or tracked vehicle used for sliding and dragging logs from the stump to a landing.
2. a machine on tracks or wheels used to drag logs from the forest to a landing where they can be further processed and loaded on a truck. Skidders are equipped with cables or grapples.
3. a machine that is used to skid or drag felled logs from the forest.
4. a machine that transports wood in the forest by supporting it at one end and dragging the other across the ground.
5. a self-propelled vehicle used to transport logs, generally by dragging them with a grapple or chokers and can be wheeled or tracked.

- **skidding**

1. the process of sliding and dragging logs from the stump to a landing, usually applied to ground-based as opposed to highlead operations.
2. dragging logs to a landing.
3. the act of moving trees from the site of felling to a loading area or landing. Skidding may be accomplished by tractors, horses, or specialised logging equipment.

- **skidding line**

see **slack-pulling line**. A line on a slack-pulling carriage that is payed out to hook to chockers. Provides lateral yarding capability and obviates the need to lower the carriage to the ground each time to hook logs.

- **skills**

great ability or proficiency; expertness that comes from training, practice, etc.

- **skyline**
  1. a type of cable logging system in which a skyline is stationary and a carriage moves along it carrying logs above the ground, from the felling site to the landing.
  2. a cable logging method in which a block or carriage rides on a wire rope called a skyline.
  3. a cableway stretched tautly between two points and used as a track for a block or carriage.

- **skyline carriage**
  see **carriage**.

- **skyline corridor**
  see **cableway**.

- **skyline logging**
  a logging method in which a block or carriage rides on a skyline.

- **skyline profile**
  the ground profile under a skyline. See **profile**.

- **skyline slope**
  the slant or inclination of the skyline chord, usually expressed in percent.

- **slackline**
  a live skyline yarding system in which the carriage is pulled to the woods by a haulback line. The skyline is lowered by slackening the line to permit the chokers to be attached to the carriage, and the turn is brought to the landing by the mainline. Lateral movement is provided by sideblocking. A slackline requires a three drum yarder.

- **slack-pulling carriage**
  a carriage capable of pulling in logs (lateral yarding) through the use of a slack-pulling line.

- **slack-pulling line**
  a line used to pull out the main or skidding line through a carriage. (see below).

- **slash**
  1. the residue left on the ground as a result of forest and other vegetation being altered by forest practices or other land use activities.
  2. tree tops, branches, bark, or other residue left on the ground after logging or other forestry operations.
  3. tree debris left after a natural catastrophe.
  4. the residue, e.g., treetops and branches, left on the ground after logging or ac-

cumulating as a result of storm, fire, girdling or delimbing.
5. branches and tops of trees left on the ground to decay and nourish the soil.
6. debris left on the ground after logging is complete.
7. branches, tops and cull trees left on the ground following a harvest. Although some of this material can be used for firewood, slash may be arranged in brush piles to provide wildlife cover. Left scattered, slash can protect seedlings and sprouts from deer browsing and reduce soil erosion.

- **slide**
  a mass movement process in which slope failure occurs along one or more slip surfaces and in which the unit generally disintegrates into a jumbled mass en route to its depositional site. A debris flow or torrent flow may occur if enough water is present in the mass.

- **slope failure**
  see **slide**.

- **slope processes**
  all processes and events by which the configuration of the slope is changed; especially processes by which rock, surficial materials and soil are transferred downslope under the dominating influence of gravity.

- **slope stability**
  susceptibility of a slope to erosion and slides.

- **slump**
  a mass movement process in which slope failure occurs on a usually curved slip surface and the unit moves downslope as an intact block, frequently rotating outward. Slumps appear as discrete block movements, often in place, whereas slides usually break up and travel downslope.

- **slurry**
  a thin watery mixture of a fine insoluble material; a mixture of liquid and solid particles so small that they remain in sus-pension instead of sinking.

- **slurry tank**
  a tank containing slurry.

- **smoke management**
  the scheduling and conducting of a prescribed burning program under

predetermined burning prescriptions and firing techniques that will minimise the adverse effects of the resulting smoke production in smoke-sensitive areas.

- **smoke-sensitive area**
  an area that has been identified in which smoke accumulations may cause a safety or public health hazard, or may unreasonably deny aesthetic enjoyment to the public.

- **smolt**
  a juvenile salmonid which has reared in-stream and is preparing to enter the ocean. Smolts exchange the spotted camouflage of the stream for the chrome of the ocean.

- **SMZ**
  Streamside Management Zone, also know as RMZ (Riparian Management Zone). Provides buffer for shade, soil stability and woody debris.

- **snag**
  1. a standing dead tree or part of a dead tree from which at least the smaller branches have fallen.
  2. any standing dead, partially dead, or defective (cull) tree at least 10 inches in diameter at breast height and at least 6 feet tall.
  3. a standing dead tree with few branches, or the standing portion of a broken-off tree. Snags may provide feeding and/or nesting sites for wildlife.

- **snag condition**
  a description of the deterioration of a standing dead tree.
  Stage 1. Branches, twigs, bark intact.
  Stage 2. Loose bark, secondary branches gone.
  Stage 3. Bole clean.
  Stage 4. Top broken.
  Stage 5. Top broken, bole decomposed.

- **snap guy**
  a guyline rigged on the same side of a standing tower yarder as the operating line(s) which prevents the tower from being pulled over backwards by the opposing guys if the operating line(s) break.

- **snatch block**
  a block that can be opened on one side to allow a cable

or rope to be laid in the block, instead of threading it through from one end.

▪ **snorkle**

a wooded or steel boom extension mounted on a loader to increase the distance that logs can be reached for loading.

▪ **snub**

to lower anything. To assist one machine down a grade by holding it back with another and connected by a 'snubbing line'.

▪ **snubbing line**

a line used in skyline logging to retard a load.

▪ **softwood**

1. a cone-bearing tree species, usually evergreen, with needles or scale-like leaves such as pine, spruce, fir and cedar.

2. a coniferous tree. The term softwood is commonly used but not strictly accurate; the wood of many conifers is harder than that of some so-called hardwood trees.

3. trees that have needle-shaped leaves; also called evergreens because they remain green throughout the year.

4. a tree belonging to the order Coniferales. Softwood trees are usually evergreen, bear cones, and have needles or scalelike leaves. They include pine, spruces, firs and cedars.

5. cone-bearing trees with needle or scale-like leaves such as Douglas-fir, western red cedar and ponderosa pine.

▪ **soil**

1. the naturally occurring, unconsolidated mineral or organic material at the surface of the earth that is capable of supporting plant growth. It extends from the surface to 15 cm below the depth at which properties produced by soil-forming processes can be detected. The soil-forming processes are an interaction between climate, living organisms, and relief acting on soil and soil parent material. Unconsolidated material includes material cemented or compacted by soil-forming processes. Soil may have water covering its surface to a depth of 60 cm or less in the driest part of the year.

2. the accumulation of mineral particles and organic matter that forms a superficial layer over large parts of Earth's surface. It provides support and nutrients for plants and is inhabited by numerous and various microorganisms and animals.

- **soil bulk density**

the mass of undisturbed or disturbed dry soil per unit bulk volume. The bulk volume is determined before drying to a constant weight at 105 c. The value is expressed in grams per cubic centimetre (g/cc).

- **soil compaction**

1. a physical change in soil properties that results in a decrease in porosity and an increase in soil bulk density and soil strength.
2. increase in the soil density that reduces soil drainage and soil aeration, often leading to decreased plant growth.

- **soil cover**

the type of cover on the soil surface:

- Soil covered by live vegetation.
- Soil covered by litter.
- Soil covered by rock.
- Soil covered by pavement.
- Soil not covered- exposed.

- **soil displacement**

the movement of the forest floor (litter, duff and humus layers) and surface soil from one place to another by mechanical forces such as a blade used in piling or windrowing. Mixing of surface soil layers by discing, chopping or bedding operation, are not considered displacement.

- **soil displacement hazard**

a soil displacement hazard as determined in accordance with procedures set out in the Ministry of Forests' publication 'Hazard Assessment Keys for Evaluating Site Sensitivity to Soil Degrading Processes Guidebook,' as amended from time to time.

- **soil disturbance**

1. disturbance caused by a forest practice on an area covered by a silviculture prescription or stand management prescription including areas occupied by excavated or bladed trails of a temporary nature, areas occupied by corduroyed

trails, compacted areas, and areas of dispersed disturbance.

2. any impact on the soil surface, including the litter and humus layers at the soil surface. Soil disturbance can be very slight (removal of the litter layer) or very heavy (rutting and soil compaction).

- **soil disturbance hazard**

an assessment of the susceptibility of a soil to adverse impacts on its productive capability due to soil compaction, soil puddling, surface erosion, mineral soil displacement, mass wasting or forest floor displacement.

- **soil drainage class**

natural soil drainage refers to the rapidity and extent of the removal of water from the soil, in relation to incoming water. This is especially true of water by surface runoff and by flow through the soil to underground spaces. Soil drainage, as a condition of the soil, refers to the frequency and duration of periods when the soil is free of saturation or partial saturation. The seven classes are:

Excessively Drained: Soils with very high and high hydraulic conductivity and low water holding capacity. They are not suited for crop production unless irrigated.

Somewhat Excessively Drained: Soils with high hydraulic conductivity and low water holding capacity. Without irrigation only a narrow range of crops can be grown and yields are low.

Well Drained: Soils with intermediate water holding capacity. They retain optimum amounts of moisture, but they are not wet close enough to the surface or long enough during the growing season to adversely affect yields.

Moderately Well Drained: Soils that are wet close enough to the surface for long enough that planting or harvesting operations are adversely affected unless artificial drainage is provided. Moderately well drained soils, commonly have a layer with low hydraulic conductivity, wet state relatively high in the profile, additions of water by seepage, or some combination of these conditions.

Somewhat Poorly Drained: Soils that are wet close enough to the surface or long enough that planting or harvesting operations are markedly restricted unless artificial drainage is provided. Somewhat poorly drained soils commonly have a layer with low hydraulic conductivity, wet state high in the profile, additions of water through seepage, or a combination of these conditions.

Poorly Drained: Soils commonly wet at or near the surface during a considerable part of the year. Poorly drained conditions are caused by a saturated zone, a layer with low hydraulic conductivity, seepage, or a combination of these conditions.

Very Poorly Drained: These soils are wet to the surface most of the time. These soils are wet enough to prevent the growth of important crops (except rice) unless artificially drained.

- **soil erosion**

wearing away of the earth's surface by water, wind or ice.

soil erosion type

factors that affect soil erosion are climate, nature of the soil, slope, vegetation and cultivation practices.

Classes of erosion by water:

Class 1: Soils that have lost some of the original A horizon but on an average are less than 25 per cent of the original A horizon or of the uppermost 20 cm of the original A horizon was less than 20 cm thick. Throughout most of the area, the thickness of the surface layer is within the normal range of variability of the non-eroded soil. Scattered spots amounting to less than 20 per cent of the area may be modified appreciably. Evidence for class 1 erosion includes:

1. a few rills,
2. accumulation of sediment at the base of slopes or in depressions,
3. scattered spots where the plow layer contains material from below the original plow layer, and
4. evidence of formation of widely spaced deep rills or shallow gullies without consistently measurable reduc-

tion in thickness or other change in properties between the rills or gullies.

Class 2: Soils that have lost on an average 25 to 75 per cent of the original A horizon or of the uppermost 20 cm if the original A horizon was less than 20 cm thick. Throughout most cultivated areas of class 2 erosion, the surface layer consists of a mixture of the original A horizon and material from below. Some areas may have intricate patterns ranging from non-eroded spots to spots where all of the original A horizon has been removed. Where the original A horizon was very thick, little or no mixing of underlying material with the original A horizon may have taken place.

Class 3: Soils that have lost on an average 75 per cent or more of the original A horizon or of the uppermost 20 cm, if the original A horizon was less than 20 cm thick. In most areas of class 3 erosion, material below the original A horizon is exposed at the surface in cultivated areas. The plow layer consists entirely or largely of material that was below the original A horizon. Even where the original A horizon was very thick, at least some mixing of underlying material with the original A horizon has generally taken place.

Class 4: Soils that have lost all of the A horizon or the uppermost 20 cm, if the original A horizon was less than 20 cm thick plus some or all of the deeper horizons throughout most of the area. The original soil can be identified only in spots. Some areas may be smooth, but most have an intricate pattern of gullies.

Classes of erosion by wind:

Class 1: Soils that have lost some of the original A horizon but on an average less than 25 per cent of the original A horizon or of the uppermost 20 cm, if the original A horizon was less than 20 cm thick. Throughout most of the area, the thickness of the surface layer is within the normal range of variability of the non-eroded soil. Scattered spots amounting to less than 20 per cent of the area may be modified appreciably. Evidence for class 1 wind erosion includes:

1. a surface layer coarser in texture than in the non-eroded areas nearby,
2. low mounds of sand, and perhaps, coarser particles, and
3. lower depressions, or swales, from which soil material has been blown.

Class 2: Soils that have lost on an average 25 to 75 per cent of the original A horizon or of the uppermost 20 cm, if the original A horizon was less than 20 cm thick. Throughout most cultivated areas of class 2 wind erosion, the surface layer consists of a mixture of the original A horizon and material from below. Most class 2 areas consist of an intricate pattern of eroded and non-eroded spots and some spots of deposition.

Class 3: Soils that have lost on an average 75 per cent or more of the original A horizon or of the uppermost 20 cm, if the original A horizon was less than 20 cm thick. In most areas of class 3 wind erosion, material below the original A horizon is exposed at the surface. In places, enough amount of the original surface layer remains such that it can be identified. The removal of material is generally not uniform. A few blowout areas may be included.

Class 4: Soils that have lost all the A horizon or the uppermost 20 cm, if the original A horizon was less than 20 cm thick plus some or all of the deeper horizons throughout most of the area. The original soil can be identified only in spots. Depressions where all of the material has been blown out are common. Areas between blowouts may be buried by soil from the blowouts.

- **soil map unit**

a named portion of a landscape shown by a closed delineation and symbol on a soil map.

- **soil pit**

an excavation into the mineral soil of sufficient depth to allow assessment of variability in soil physical properties within a defined area of land.

- **soil puddling**

  a physical change in soil properties due to shearing forces that alters soil structure and porosity. Puddling occurs when the soil is at or near liquid limit.

- **soil structure**

  structure is described by grade, class and type. Terms are used to describe natural aggregates in the soil called peds in contrast to clods caused by disturbance, fragments by rupture of peds, and concentrations by local concentrations of compounds that irreversibly cement the soil grains together. Classes are as follows:

  - Granular, approximately spherical with no accommodation of faces to surrounding peds.
  - Platy, with vertical dimension small with regard to horizontal dimensions; faces accommodate with those of adjacent peds.
  - Prismatic, without rounded caps, vertical faces well defined and with angular vertical length relatively long with respect to horizontal dimensions; faces accommodate with those of adjacent peds.
  - Columnar, with rounded caps, otherwise similar to the prismatic.
  - Angular blocky, block like with all 3 dimensions of the same order of magnitude, faces flattened, most vertices sharply angular; faces accommodate with those of adjacent peds.
  - Subangular blocky, similar to angular blocky but both rounded and flattened faces occur with many rounded vertices.
  - Structureless: massive or single grain.

- **soil texture**

  1. the relative proportions of clay, silt, and sand (less than 2mm in diameter). Clay particles are the smallest, silt particles are intermediate, and sand particles are the largest. Loams contain various mixtures of the three basic particle sizes. Rock fragments in the soil can modify textural names depending on size and amount. Stones and boulders on the surface can affect use and coverage should be estimated.

2. the feel or composition of the soil (sand, silt, or clay) as determined by the size of the soil particles.
3. the relative proportion of the various size groups of individual soil particles.

- **soil type**
soils that are alike in all characteristics, including texture of the topsoil. Soil maps and information on site index, erodibility and other limiting properties are available from your county Soil Conservation Service offices.

- **spacing**
1. the removal of undesirable trees within a young stand to control stocking, to maintain or improve growth, to increase wood quality and value, or to achieve other resource management objectives.
2. removing trees from a stand to decrease the stand density, to distribute the crop of trees more evenly over the growing room

- **span**
the horizontal distance between two adjacent skyline supports.

- **spar**
a tree, wood mast, or metal tower used to support rigging for one of the many cable yarding systems. In the U.S. today, most commonly a mobile, portable tower is the rule. See **tower**.

- **spasmoneme**
a contractile element in the stalk of some peritrich ciliates.

- **species**
1. a singular or plural term for a population or series of populations of organisms that are capable of interbreeding freely with each other but not with members of other species. Includes a number of cases:
a) endemic species: a species originating in, or belonging to, a particular region. Both 'endemic' and 'indigenous' are preferred over 'native'.
b) exotic species: a species introduced accidentally or intentionally to a region beyond its natural range. 'Exotic' is preferred over 'alien', 'foreign' and 'non-native.'
c) a group of related organisms with common characteristics that are capable of

interbreeding. Loblolly and Virginia pine are examples of two different tree species.
d) a category of biological ranking just below the genus or subgenus category. Members of the same species are closely related organisms that are potentially able to breed with one another.
e) a subordinate classification to a genus; reproductively isolated organisms that have common characteristics, such as eastern white pine or white-tailed deer.

- **subspecies**
  a subdivision of a species. A population or series of populations occupying a discrete range and differing genetically from other subspecies of the same species.

- **species at risk**
  any wildlife species that, in the opinion of the Deputy Minister of Environment, Lands and Parks, or a person authorised by that deputy minister, is threatened, endangered, sensitive or vulnerable,

- **species composition**
  the percentage of each recognised tree species comprising the forest type based upon the gross volume, the relative number of stems per hectare or basal area.

- **species conversion**
  a change from one tree species to another.

- **species diversity**
  the number of different species and their relative abundance in a given area.

- **species richness**
  the number of species present in a community or a defined area.

- **spicule**
  delicate pointed structures lying external to the body and usually pointed away from it. Like spines, but invariably excreted and more delicate.

- **spore**
  1. a specialised structure consisting of one or few cells and serving any or all of the following three functions: reproduction, dissemination, survival.
  2. unicellular body produced by plants, fungi, and some microorganisms. Often a resting stage, encased in a protective coat, adapted to

resist heat, desiccation, or other unfavourable environmental conditions. Spores can give rise to a new individual either directly or after fusion with another spore.

- **sporophore**
  a spore-producing or supporting structure.
- **sporulation**
  the process of spore development.
- **spot burning**
  a modified form of broadcast burning in which only the larger accumulations of slash are fired and the fire is confined to these spots.
- **spot landing**
  a round or rectangular area cleared for use as a landing.
- **spring**
  a flow of ground water emerging naturally onto the earth's surface and used as a domestic water source within a community watershed. The watershed area of a spring is defined as the total recharge area of the spring.
- **spring seep**
  a class of wetland created by groundwater emerging on lower slopes in small pools surrounded by vegetation. These create snow-free zones critical for wildlife feeding during winter.
- **springwood**
  the first-formed woody portion of the annual growth ring of a shrub or tree.
- **spruce budworm**
  an insect that damages conifer trees by defoliation.
- **spur road**
  a short, low standard branch road generally accessing one or two landings; often dead ending at the last landing.
- **stabilised road width**
  the width of the travelled portion of the road that has been surfaced with material of sufficient strength and quantity to support the intended traffic.
- **stadium**
  the period of time between two successive molts.
- **stage**
  one of the successive principal divisions in the life cycle of an insect, e.g., egg, nymph, larva, prepupa, pupa, adult.

- **stage-felling**
  to fell timber in several successive cuts to reduce bole breakage and damage to understory reproduction. AKA 'stage cutting'.

- **stagged pants**
  work pants shortened to prevent limbs catching the cuff and tripping the logger. Often left unhemmed and frayed at the bottom.

- **stag head**
  death of limbs and main branches of a tree in the upper crown, giving the appearance of antlers.

- **stagnant**
  of stands whose growth and development have all but ceased due to poor site and/or excessive stocking.

- **stalk**
  a support which connects a leaf or a bud to a twig.

- **stamen**
  the male part of a flower that makes pollen.

- **stand**
  1. a community of trees sufficiently uniform in species composition, age, arrangement, and condition to be distinguishable as a group from the forest or other growth on the adjoining area, and thus forming a silviculture or management entity.
  2. an easily defined area of the forest that is relatively uniform in species composition or age and can be managed as a single unit.
  3. a group of trees, occupying a specific area, that are fairly uniform in species composition, age and condition.
  4. a contiguous group of trees sufficiently uniform in species composition, arrangement of age classes and condition to be considered a unit.
  5. a grouping of vegetation sufficiently uniform in species composition, age, and condition to be distinguished from surrounding vegetation types and managed as a single unit.

- **stand age**
  the mean age of the dominant and codominant trees in the stand.

- **stand composition**
  the proportion of each tree species in a stand expressed as a percentage of either the total number, basal area or

volume of all tree species in the stand.

- **stand condition**

a classification of forest stands based upon the age of maturity and structure of the overstory and understory.

Old-Growth Stands: Ecosystems distinguished by old trees and related structural attributes. Old growth encompasses the later stages of stand development which typically differ from earlier stages in a variety of characteristics that may include tree size, accumulations of large dead woody material, number of canopy layers, species composition, and ecosystem function. The age at which old growth develops and the specific structural attributes that characterise old growth will vary widely according to forest type, climate, site conditions and disturbance regime. For example, old growth in fire-dependent forest types may not differ from younger forests in the number of canopy layers or accumulation of down woody material. However, old growth is typically distinguished from younger growth by several of the following structural attributes: a) Large trees for species and site. b) Wide variation in tree sizes and spacing. c) Accumulations of large-size dead standing and fallen trees that are high relative to earlier stages. d) Decadence in the form of broken or deformed tops or bole and root decay. e) Multiple canopy layers. f) Canopy gaps and understory patchiness.

Young-Growth Stand: Any forested stand not meeting the definition of old growth.

- **stand conversion**

changing the species composition of a stand to more desirable tree species which are less susceptible to damage or mortality from certain insects or diseases.

- **stand density**

a relative measure of the amount of stocking on a forest area. Often described in terms of stems per hectare.

- **stand development**

the part of stand dynamics concerned with changes in stand structure over time.

stand dynamics
the study of changes in forest stand structure over time, including stand behavior during and after disturbances.

stand history
the kind of disturbance (prior to plot establishment) on the sample location.

stand level
the level of forest management at which a relatively homogeneous land unit can be managed under a single prescription, or set of treatments, to meet well-defined objectives.

stand management prescription
a site-specific operational plan describing the nature and extend of silviculture activities planned for a free growing stand of trees to facilitate the achievement of specified or identified social, economic and environmental objectives.

stand model
a computer model that forecasts the development of a forest stand, usually in terms of stand attributes such as mean diameter or height.

stand origin
the apparent source of vegetation on the location-natural or artificial regeneration.

stand size class
a classification of land based on the stocking of all live vegetation of various sizes.
Nonstocked with vegetation: Land less than 10-percent stocked with trees, but having vegetation.
Grass-Forb Stands: Stands less than 10 percent stocked with trees. Shrubs less than 40 percent crown canopy.
Shrub Stands: Stands less than 10 percent stocked with trees. Shrubs greater than 40 percent crown canopy.
Seedling-Sapling Stands: Stands at least 10 percent stocked with live trees of all sizes, of which half or more of the stocking consists of seedling and/or saplings (trees < 5.0" d.b.h.).
Poletimber Stands: Stands at least 10 percent stocked with live trees of which half or more of the stocking is in trees, 5.0 inches d.b.h. and larger, and in which the stocking of poletimber (softwoods-5.0 to 8.9 inches

d.b.h., hardwoods-5.0 to 10.9 inches d.b.h.) exceeds the stocking of sawtimber (trees larger than poletimber).

Sawtimber Stands: Stands at least 10-percent stocked with live trees of which half or more of the stocking is in trees 5.0 inches d.b.h. and larger, in which the stocking of sawtimber trees is at east equal to the stocking of poletimber trees.

- **stand strategy**

  a documented plan of stand treatments to achieve management objectives during the life of a particular stand.

- **stand structure**

  1. the distribution of trees in a stand, which can be described by species, vertical or horizontal spatial patterns, size of trees or tree parts, age or a combination of these.

  2. a description of the distribution and representation of stand age and stand size classes within a stand.

  Even-aged single storied: Theoretically, stands in which all the trees are one age. In actual practice, these stands are marked by an even canopy of uniform height characterised by intimate competition between trees of approximately the same size. The smaller trees are usually tall spindly members of the stand that have fallen behind their associates. The greatest number of stems are in a diameter class represented by the average of the stand; there are fewer trees in the classes above and below this mean. A single even canopy characterises the stand. The greatest number of trees are in a height class represented by the average height of the stand; there are substantially fewer trees in height classes above and below this mean. The ages of the trees usually do not differ by more than 20 years.

  Even-aged two-storied: Stands composed of two distinct canopy layers, such as, an overstory and understory sapling layer possibly from seed tree and shelterwood operations. This may also be true in older plantations where tolerant hardwoods may become established as management intensity decreases (burning and other means of understory control). Two relatively even canopy levels can be recognised in the stand. The frequency distribution of trees by height class tends to be bimodal. Understory or overtopped trees are common. Neither canopy level is necessarily continuous or closed, but both canopy levels tend to be uniformly distributed across the stand. The average age of each level differs significantly from the other.

Uneven-aged: Theoretically, these stands contain trees of every age on a continuum from seedlings to mature canopy trees. In practice, uneven-aged stands are characterised by a broken or uneven canopy layer. Usually the largest number of trees is in the smaller diameter classes. As trees increase in diameter, their numbers diminish throughout the stand. Many times, instead of producing a negative exponential distribution of diminishing larger diameters, uneven-aged stands behave irregularly with waves of reproduction and mortality. Consider any stand with 3 or more structural layers as uneven-aged. Logging disturbance (examples are selection, diameter limit and salvage cutting) will give a stand an uneven-aged structure.

Mosaic: At least two distinct size classes are represented and these are not uniformly distributed, but are grouped in small repeating aggregations, or occur as stringers less than 120 feet wide, throughout the stand. Each size class aggregation is too small to be recognised and mapped as an individual stand. The aggregations may or may not be even-aged.

- **stand table**

  a summary table showing the number of trees per unit area by species and diameter class, for a stand or type. The data may also be presented in the form of a frequency distribution of diameter classes.

- **stand tending**

  a variety of forest management treatments, including spacing, fertilisation, pruning, and commercial thinning, carried out at different stages during a stand's development.

- **stand year of origin**
  year the stand was planted or created.

- **standard**
  the required level or measure of practice established by authority of the forest practices code and referenced in legislation.

- **standing**
  status held by a person or group which allows the person or group to challenge or appeal a particular decision.

- **standing skyline**
  a fixed skyline not running during logging operations; for example, a skyline anchored at one or both ends. The skyline length is fixed during yarding and is not raised or lowered during the cycles. Not used very often in today's logging. Preference is given to live skylines or running skylines.

- **starch**
  sugar is stored in a tree as starch, which is a complex carbohydrate. This is an available energy source.

- **starch grain**
  a term used loosely to refer to refractile masses of polysaccharides which are accumulated as storage products in the cell.

- **station**
  a 100 ft long segment or distance of a road grade or skyline profile.

- **statutory framework**
  where forest practices are primarily regulated by legislation.

- **sterile tree**
  trees incapable of reproducing.

- **stewardship**
  1. caring for land and associated resources and passing healthy ecosystems to future generations.
  2. the concept of responsible caretaking; the concept is based on the premise that we do not own resources, but are managers of resources and are responsible to future generations for their condition.
  3. the wise management and use of forest resources to en-

sure their health and productivity for the future with regard for generations to come.

- **stewardship forest**
  a privately owned forest tract that exhibits integrated forest management to protect and enhance wildlife, timber, recreation, natural beauty and soil and water quality.

- **Stewardship Incentive Program (SIP)**
  a cost-sharing program available to forest landowners who have a multiresource forest stewardship plan. Practices include cost-sharing assistance for the enhancement of forest recreation, fisheries, wildlife, and timber production and the protection of soil and water, wetlands, riparian zones and rare and endangered species.

- **stigma**
  the sticky surface of a flower pistil on which pollen adheres during pollination.

- **stipple**
  pigmented spots up to a few millimeters in diameter, often on the upper surface of leaves.

- **stocking**
  1. a measure of the area occupied by trees, usually measured in terms of well- spaced trees per hectare, or basal area per hectare, relative to an optimum or desired level.
  2. a description of the number of trees, basal area, or volume per acre in a forest stand compared with a desired level for balanced health and growth. Most often used in comparative expressions, such as well-stocked, poorly stocked, or overstocked.
  3. a relative number of trees or volume per acre. A forest stand is most often described as being well-stocked, poorly-stocked, or over-stocked.

- **stocking class**
  a numeric code representing a range of stems per hectare, sometimes estimated by crown closure on aerial photographs, e.g. stocking class 1 is mature with 76+ stems/ha of > 27.5 cm DBH; class 2 is mature with <76 stems/ha; class 0 is immature.

- **stocking percent**
  the amount of live trees on a given area in relation to what

is considered the optimum. A calculation using either the total number of trees, total basal area, or total volume per unit area divided by the optimum total number of trees, optimum total basal area or optimum total volume for a particular species and management objective, expressed as a percent. The classes are 10 percent intervals with an overall range from non-stocked to 100% stocking, relative stocking basis. Data in the form of absolute stocking (0-167 percent basis) should be reduced to relative stocking by dividing by a factor of 1.67.

- **stocking plan**
  a plan that provides objectives and strategies for land allocation and/or resource management, including regional plans sub regional plans and local resource plans.

- **stocking standard**
  the required range of healthy, well-spaced, acceptable trees.

- **stocking survey**
  the determination of the stocking of an area of both well-spaced and total trees; also used to generate an inventory label.

- **stocks**
  groups of fish that migrate together are called runs, or stocks. The threatened or endangered species listing usually refers to stocks on specific rivers (e.g. winter-run Chinook on Sacramento River).

- **stoma**
  1. a small opening found in the epidermal layer of plants, allowing access for carbon dioxide and the release of water. Stomata are surrounded by guard cells that control the opening size.
  2. a breathing pore on a leaf.

- **stomata**
  openings in a leaf through which gases and water enter or leave.

- **stomatocyst**
  a type of cyst, with a siliceous wall and a single plugged opening, formed by some chrysomonads (stramenopiles).

- **strand**
  a component of wire rope consisting of wires spirally wound together which is then helically laid around the core to form the rope. See **wire rope**.

- **strap**
  a short cable with an eye in each end. AKA 'sling'.

- **strategic land-use planning**
  planning at the regional, sub-regional and, in some cases, at the local level which results in land allocation and/or resource management direction. Strategic land-use planning at the regional and sub-regional level involves the preparation of resource management zones, objectives and strategies.

- **strategic planning**
  long range planning which allows an individual or organisation to exercise control over its future and achieve a sustainable competitive advantage in the marketplace. Without an overall strategy, short-term tactics are often misguided.

- **strategy**
  a broad non-specific statement of an approach to accomplishing desired goals and objectives.

- **straw drum**
  a small drum on a yarder holding the straw line.

- **straw line**
  a light cable, usually 3/8 inch or less in diameter, used to string heavier lines. See **haywire**.

- **stream**
  1. a watercourse, having an alluvial sediment bed, formed when water flows on a perennial or intermittent basis between continuous definable banks.
  2. a water course that has ordinary high watermark, has beds and banks, flows at least periodically, has an easily identifiable beginning and end an does not loses it character as a water course even though it may break up and disappear temporarily and reappear down stream.

- **stream bank**
  the rising ground bordering a stream channel.

- **stream channel**
  the streambed and banks formed by fluvial processes,

including deposited organic debris.

- **stream channel-bank angle**
  a measure of the angle formed by the downward sloping streambank as it meets the more horizontal stream bottom.

- **stream culvert**
  a culvert used to carry stream flow in an ephemeral or perennial stream channel from one side of the road to the other.

- **stream gradient**
  the general slope, or rate of vertical drop per unit of length of a flowing stream.

- **stream order**
  a numbering scheme used to characterise the relative position of stream channels within a drainage. Given a map, the first-order streams are those which have no tributaries. The second-order streams are those which have as tributaries only first-order channels. The third-order stream is formed when two second-order channels come together, etc.

- **stream shade cover**
  the percent of shade covering a stream or lake shore water surface averaged for the day.

- **stream type**
  alpha-numeric identification given to reoccurring stream channel types based on measurable morphological features.

- **stream width**
  the width of streams or rivers. Stream or River-means a natural water course.

- **streambank undercut**
  a measure of the furthest point of protrusion of the bank to the furthest undercut of the bank. Streambank undercut provides cover for fish and may be conducive to producing high biomass of fish. Undercut is a good indicator of how successfully streambank are protected under alternative and uses, such as livestock grazing and road building.

- **streambed**
  the bottom of the stream below the usual water surface.

streamflow

measure of the volume of water passing a given point in a stream channel at a given point in time. Streamflow is a function of depth, width and velocity of water in a channel. Changes in streamflow affect the available habitat for fish spawning or rearing. Streamflow can be determined using a number of methods. Expressed in cubic feet/sec. or cubic meters/sec.

Streamside Management Area (SMA)

(also commonly referred to as a riparian area) Land and vegetation areas next to lakes and streams where management practices are modified to protect water quality, fish and other aquatic resources. These areas are complex ecosystems that provide food, habitat and movement corridors for both aquatic and terrestrial communities as well as helping to minimise nonpoint source pollution impacts to surface water.

Streamside Management Zone (SMZ)

1. the land, together with the vegetation that supports it, immediately in contact with the stream and sufficiently close to have a major influence on the total ecological character and functional processes of the stream.

2. an area adjacent to a stream in which vegetation is maintained or managed to protect water quality. The width depends on slope, but 50 feet is the normal minimum. Trees may be removed from SMZs as long as the stream bed is not disrupted and sufficient vegetation is left to protect water quality.

3. a designated area that consists of the stream itself and an adjacent area of varying width where management practices that might affect water quality, fish, or other aquatic resources are modified. The SMZ is an area of closely managed activity, not an area of exclusion.

4. vegetative areas adjacent to streams, rivers, lakes and

ponds managed for the protection of water quality stream habitat.

- **strip**
  the area in which a set of cutters works a cutting strip.

- **structural panel**
  panels such as plywood and OSB that are used in applications where strength and stiffness are required, e.g., roofs, walls, floors etc.

- **stump**
  the lower end of a tree remaining in the ground after most of the stem or trunk has been removed

- **stump height**
  the vertical distance from the ground on the uphill side to the top of the stump on cut trees. Vertical distance from the ground to a stump height set by study objectives or local utilisation practice for uncut trees.

- **stump pull**
  splinters of wood pulled from the butt log. If the splinters are too long, stump pull is considered excessive and devalues the butt log.

- **stumpage**
  1. is the fee that individuals and firms are required to pay to the government when they harvest Crown timber in British Columbia. Stumpage is determined through a complex appraisal of each stand or area of trees that will be harvested for a given timber mark. A stumpage rate ($ per m3) is determined and applied to the volume of timber that is cut (m3). Invoices are then sent to individuals or firms.
  2. the value or volume of a tree or group of trees as they stand uncut in the woods (on the stump).
  3. the price that must be paid to the provincial government for timber harvested on Crown land.
  4. the value of timber as it stands uncut. The residual value after all logging costs are taken from the delivered price of logs at the mill yard.
  5. the commercial value of standing trees.

- **style**
  extending from the center of a flower, it supports the

stigma where pollen adheres during pollination.

- **stylet**
  a small, stiff, needlelike tube inserted into a food source to obtain liquid food.

- **subgrade**
  the material movement necessary to construct the roadway, excluding surfacing.

- **substantial equivalence**
  a principle for safety assessment that compares a traditional non-modified product with a long history of safe use to a genetically modified product. It is substantially equivalent if it has all the characteristics of the non-modified product.

- **substrate**
  the bottom material of a lake, stream, or ocean such as mud, sand or gravel.

- **substructure**
  the part of a bridge that supports the superstructure and carries all the applied lateral and vertical loads; includes caps, sills, piles, and posts, each comprising elements known as abutments and piers.

- **subsurface drainage**
  water flow through permeable soil or rock beneath the surface of the land.

- **sub-unit plan**
  the fourth level of planning in the ministry of forests hierarchical planning system. The aggregation of a number of courses of action in map and written form designed to achieve sub-unit objectives. Normally centred on watersheds.

- **succession**
  1. the gradual supplanting of one community of plants by another, the sequence of communities being termed a sere and each stage seral. 2. the natural sequence of plant community replacement beginning with bare ground and resulting in a final, stable community in which a climax forest is reached. Foresters, wildlife biologists, and farmers constantly battle ecological succession to try to maintain a particular vegetative cover. 3. the natural series of replacements of one plant community (and the associated fauna) by another over

time and in the absence of disturbance.
4. the replacement of one plant and/or animal species over time by another in progressive development toward climax vegetation.

- **successional disking or mowing**
  a wildlife enhancement practice in which a disk harrow or rotary mower is used to knock down existing vegetation every 1 to 3 years to promote the regrowth of annuals, legumes, forbes and perennials.

- **sucking insects**
  insects that insert their mouthparts into plant tissues and withdraw nutrients and fluids through stylets.

- **suitability mapping**
  a habitat interpretation that describes the current potential of a habitat to support a species. Habitat potential is reflected by the present habitat condition or successional stage.

- **sulphate**
  a measure of sulphate concentration. Expressed as mg/l of SO4.

- **summerwood**
  the last-formed woody portion of the annual growth ring of a shrub or tree, usually containing more fibres and fewer vessels than springwood.

- **super trees**
  also referred to as 'plus trees', these are trees that have been selected for their superior traits (i.e. volume or form) to be tested and possibly utilised in a genetic tree improvement program.

- **superstructure**
  the part of a bridge found above or supported by the caps or sills, including the deck, girders, stringers and curbs.

- **supply block**
  an area of Crown land that is relatively homogeneous with respect to forest characteristics, access development and management concerns. Supply blocks are the next smaller timber management unit within a Timber Supply Area.

- **suppressed trees**
  trees or shrubs with crowns receiving no direct light either from above or from the

sides, and that will not respond to release. Usually crowns are entirely below the general level of the canopy.

- **surface erosion**
the detachment and transport of individual soil particles by wind, water or gravity.

- **surface fire**
a fire burning along the surface without significant movement into the understory or overstory, with flame length usually below 1 meter.

- **surface soil erosion**
means for an area where a forest practice has been carried out, the movement of soil particles from the area by wind, gravity or water at a rate that is greater than that which would have occurred had the forest practice not been carried out.

- **surface water**
precipitation that does not soak into the ground or return to the atmosphere by evaporation.

- **surplus forest**
a forest in which existing stands can provide more harvest volume than is needed to maintain the harvest at the level of long run sustained yield until the stands created when the existing stands are cut become available for harvest. See also **deficit forest**.

- **suscept**
any organism that can be attacked by a biotic pathogen.

- **suspended sediment**
sediment which remains in suspension in water for a considerable period of time without contact with the bottom. Sediment content is measured in the Laboratory and reported as parts per million or milligrams per litre.

- **suspension feeding**
feeding on suspended particles. The most usual ploy is filter-feeding, but not all suspension feeders feed in this way.

- **sustainability**
a state or process that can be maintained indefinitely. The principles of sustainability integrate three closely interlined ele-

ments—the environment, the economy and the social system—into a system that can be maintained in a healthy state indefinitely.

- **sustainability**
to keep in existence, to maintain or prolong

- **sustainable development**
preservation and protection of diverse ecosystems-the soil, plants, animals, insects and fungi while maintaining the forest's productivity.

- **sustainable forest development**
the development of forests to meet current needs without prejudice to their future productivity, ecological diversity, or capacity of regeneration.

- **sustainable forest management**
management regimes applied to forest land which maintain the productive and renewal capacities as well as the genetic, species and ecological diversity of forest ecosystems.

- **sustainable forestry**
forests that are sensitively managed and harvested so that they will be around for generations to come.

- **sustainable harvest**
a forest management policy in which the amount of trees harvested is not to exceed the amount of trees grown by the forest each year.

- **sustained harvest**
a level of harvest that does not exceed growth, so that at least as much is available for harvest in 50 years as today.

- **sustained yield**
1. a method of forest management that calls for an approximate balance between net growth and amount harvested.
2. management of forestland to produce a relatively constant amount of wood products, revenue or wildlife.
3. historically, a timber management concept in which the volume of wood removed is equal to growth within the total forest. The concept is applicable to nontimber forest values as well.

- **swaged**
to press on, such as a ferrule on a choker.

- **swamp**

  to clear brush and obstructions for a cat clearing roads or skidtrails. The person doing this hand work is called a swamper.

- **swede saw**

  a thin kerfed manual saw used on bygone logging operations primarily for delimbing or felling smaller diameter trees

- **swing**

  1. to move logs to a landing from a distance deck to which they have previously been yarded. AKA 'shuttle logging'.

  2. the capability of a yarder to swing to either side to position logs beside the machine.

- **swing yarder**

  any yarder that swings on a turntable, as opposed to a standing spar yarder.

- **switch back**

  a horizontal road curve used for surmounting the grade of a step hill, usually with a small radius (15-10 m) and curving 180 degrees.

- **swivel**

  a fastening device that allows the parts attached to it to rotate freely.

- **symbiosis**

  1. the living together in close association of two or more dissimilar organisms: includes parasitism, mutualism, and neutralism.

  2. living in association with another organism, normally to the mutual advantage of both or to the advantage of one - the other being unaffected. Where there is a notable discrepancy in size, the term 'symbiont' is used to refer to the smaller member of the association which may occur inside (endosymbiont) or on the outside surface (ectosymbiont) of the larger member 'host'.

- **symplesiomorphy**

  a shared ancestral character state (plesiomorphy). A trait that is shared by two or more groups due to inheritance from a distant common ancestor. Symplesiomorphies are common to all the descendants of a distant common ancestor, and they cannot be used to infer close evolutionary relationships between subsets of these descendants.

- **symptom**
visible or measurable manifestation that an organism is diseased; a change in the organism itself.

- **synapomorphy**
a shared derived character state (apomorphy). A novel evolutionary trait that is shared by two or more groups due to inheritance from an immediate common ancestor. Synapomorphies are used by phylogenetic biologists to infer close evolutionary relationships between organisms.

- **syncytium**
multinucleated mass of cytoplasm.

- **system closure**
the concept of reducing pollution at source by developing efficient ways to re-use more water and other elements in the waste stream.

- **system road**
a permanent road required for long-term management of the forest.

- **tagline**
1. a short line added to another line to lengthen it.
2. a line used to position a loading grapple. See **drop line**.

- **tailblock**
a block at the back end of the yarding area which is used to guide the haulback line and is attached to an anchor or spar tree.

- **tailhold**
the anchorage at outer end of the skyline, away from the landing. A line securing a tailblock to a stump.

- **tailspar**
a spar at the outer end of a skyline system, away from the landing, which elevates and supports one end of the skyline. AKA 'back spar'.

- **tailtree**
a standing tree used as tailspar.

- **talkie-tooter**
a portable radio transmitter, carried by one of the choker setters or rigging slinger on a cable operation, which operates a whistle on the yarder and signals the yarder engi-

neer what cable manoeuvres to initiate. Talkie-tooter is a brand name for the most commonly used make of these radio controlled signalers and is often used generically, much like Kleenex or Coke, in woods conversations. AKA a 'bug'.

- **tall oil**
  a paper making by-product commonly used to make soaps, paints, varnishes, printing inks and waterproofing agents.

- **tannins**
  pigments in plants responsible for brown colours.

- **target stocking standards**
  the number of well-spaced acceptable trees per hectare that will, in normal circumstances, produce an optimum free-growing crop; the standards at which silviculture treatments are aimed.

- **taxon**
  any named group of organisms.

- **teamwork**
  joint action by a group of people, in which individual interests are subordinated to group unity and efficiency; coordinated effort

- **telial stage (telium)**
  a stage of the rust fungi; a fruiting structure usually appearing as fine, hairlike projections from lower surfaces of infected leaves.

- **temperature, ambient**
  the hourly air temperature of the surrounding area.

- **temperature, water**
  a measure of water temperature heterogeneity.

- **temporary access structure**
  a structure that would be a permanent access structure except that it is not shown on a forest development plan, access management plan, logging plan, road permit or silviculture prescription as remaining operational after the completion of timber harvesting activities.

- **temporary bridge**
  a bridge having most of its major components constructed of untreated wood.

- **temporary roads**
  these are the most common type of woodland roads. They are designed and con-

structed for short-term use during a specific project such as timber harvesting. These roads are used only when the ground is frozen or firm. When the project is done, the temporary road is closed, all stream crossing structures are removed and the road is naturally or artificially revegetated.

- **temporary tenures**
  non-alienated lands on which the timber is alienated to private interests, but where the Crown retains ownership of the lands.

- **tending**
  any operation carried out for the benefit of a forest crop or an individual thereof, at any stage of its life. It includes operations both on the crop itself and on competing vegetation but not site preparation or regeneration cuttings.

- **tenure**
  the holding, particularly as to manner or term (i.e., period of time), of a property.

- **terminal taxa**
  the taxa or named groups at the tips of the branches of a tree.

- **terrain**
  the physical features of a tract of land.

- **terrain hazard assessment**
  an assessment or characterisation of unstable or potentially unstable slopes on forested lands. A determination of the relative potential of landslide initiation and the type of landslide that may occur on different types of terrain, based on the data obtained from a review of available maps, photos, site data and field observations.

- **terrain stability risk**
  a combined assessment of both the likelihood of landslide initiation and an order of magnitude estimate of the amount of landslide debris that might enter a stream or of the potential lengths of scour of a stream by a landslide.

- **terrestrial**
  pertaining to land - as opposed to the water (aquatic) or the air (aerial).

- **theca**
  a layer which encloses a body, may refer to a closely

adpressed rigid wall, or to a more loosely-attached rigid lorica or test or even to a soft enclosing sheath of material.

- **theront**

a form adopted by some species of ciliates that typically does not feed but moves quickly. May be thought of as an adaptation in response to a lack of food and the task of which is to hunt out new sources of food.

- **thinning**

1. a cutting made in an immature crop or stand primarily to accelerate diameter increment but also, by suitable selection, to improve the average form of the trees that remain.
2. a practice in which certain trees are removed from a dense stand to allow the remaining trees adequate sunlight, nutrients and moisture to grow at an even rate.
3. a tree removal practice that reduces tree density and competition between trees in a stand. Thinning concentrates growth on fewer, high-quality trees, provides periodic income, and generally enhances tree vigor. Heavy thinning can benefit wildlife through the increased growth of ground vegetation.
4. the process of removing excess and poorer quality trees from a stand for the purpose of improving the growth and value of the remaining crop trees.
5. removing excess and poorer quality trees from a stand for the purpose of improving the growth and value of the remaining crop of trees.
6. removal of trees to encourage growth of other selected individual trees. May be commercial or precommercial.

- **thoracic**

belonging or attached to the thorax.

- **thorax**

the body region behind the head, which bears wings and true (jointed) legs if present.

- **thousand board feet**

unit of measurement equal to 1,000 feet of wood having a thickness of 1 inch.

- **threatened or endangered plant communities**

ecosystems, as listed by the Conservation Data Centre, that:

are restricted in their distribution over a natural landscape (e.g., freshwater wetlands within certain biogeoclimatic) or are restricted to a specific geographic area or a particular type of local environment; or

were previously widespread or common but now occur over a much smaller area due to extensive disturbance or complete destruction by such practices as intensive harvesting or grazing by introduced species, hydro projects, dyking and agricultural conversion.

- **threatened or endangered species**

1. threatened or endangered species as identified by the Ministry of Environment, Lands and Parks. The taxa on the red list are either Extirpated, Endangered or Threatened, or are being considered for such status. Any indigenous taxon (species or sub-species) threatened with imminent extinction or extirpation throughout all or a significant portion of its range in British Columbia is Endangered. Threatened taxa are those indigenous species or subspecies that are likely to become endangered in British Columbia if factors are nor reversed.

2. species that are likely to become endangered in the near future.

3. a species likely to become endangered in the foreseeable future, throughout all or a significant portion of its range, unless protected.

- **tieback**

a line attached to a secondary stump from the main anchor stump to distribute the load on the anchor. Commonly used where one large stump is not available to provide adequate anchorage.

- **tightlining**

1. In highlead logging, a method of lifting the logs or butt rigging over obstructions by tightening the haulback line with brakes while pulling them in a line.

2. Moving rigging from one road to another by changing the tailblocks and pulling the cable to the new road by tightening the mainline.

- **timber**

1. trees, whether standing, fallen, living, dead, limbed, bucked or peeled.
2. general term applied to forests and their products. Sawed lumber more than 4 by 4 inches in breadth and thickness.
3. wood.

- **timber appraisal**

economic appraisal of the monetary value of a timber stand.

- **timber cruising**

1. the collection of field data on forests commonly by the measurement and recording of information in sample plots. Includes the measurement and estimation of volumes of standing trees.
2. the process of estimating the quality, quantity, and characteristics of trees in a forest.

- **timber harvesting land base**

the portion of the total area of a management unit considered to contribute to, and be available for, long-term timber supply. The harvesting land base is defined by reducing the total land base according to specified management assumptions.

- **timber licence**

area-based tenures which revert to the government when merchantable timber on the area has been harvested and the land reforested. Many of these licences have been incorporated into tree farm licences.

- **timber management prescriptions**

recommended forest management practices, usually pertaining to the sub-unit and operational levels of planning.

- **timber mark**

a hammer indentation made on cut timber for identification purposes.

- **timber operability**

in a planning context, the term refers to the economic suitability of timber for harvesting. Parameters to consider in assessing operability include: terrain, timber quality, timber size, operating season, labour costs, development costs and transportation costs. In the Environmental Protection Area program, operability refers to freedom from harvesting constraints which include environmental protection and other forest uses.

- **Timber Products Output (TPO)**
timber products cut from roundwood and by-products of wood-manufacturing plants. Roundwood products include logs, bolts or other round sections cut from growing stock trees, cull trees, salvable dead trees, trees on nonforest land, noncommercial species, sapling-size trees, and limbwood. By-products from primary manufacturing plants include slabs, edging, trimmings, miscuts, sawdust, shavings, veneer cores and clippings, and screenings of pulp mills that are used as pulp chips or other products.

- **Timber Stand Improvement (TSI)**
1. improving the quality of a forest stand by removing or deadening undesirable species to achieve desired stocking and species composition. TSI practices include applying herbicides, burning, girdling, or cutting.
2. improving the quality of a forest stand by removing or deadening undesirable species to achieve desired stocking and species composition.
3. a combination of intermediate treatments designed to improve growth and composition of the forest; often spoken of as TSI.

- **timber supply**
the available timber categorised by species, end-use and relative value.

- **timber supply analysis**
an assessment of future timber supplies over long planning horizons (more than 200 years) by using timber supply models for different scenarios identified in the planning process.

- **timber supply model**
an analytical model (usually computer-based) that simulates the harvest and growth of collections of forest stands over several de-

cades according to specific data and management assumptions.

- **timber treatment opportunity class**
  a class to identify the physical opportunity for increasing timber production.

- **timber trees**
  the forest crops and stands containing trees of commercial size and quality suitable for sawing into lumber.

- **timber utilisation**
  the dimensions and quality of timber that is actually cut and removed from an area.

- **timberland**
  forestland producing or capable of producing crops of industrial wood (more than 20 cubic feet per acre per year), and not withdrawn from timber utilisation. Formerly known as commercial forestland.

- **time since disturbance**
  the number of years between when the most recent disturbance took place and the time of plot establishment.

- **tissue culture**
  the cultivation of cells or tissues in an artificial environment, such as the laboratory or in vitro.

- **thermo-mechanical pulp**
  a pulp produced using steam, heat and enormous pressure to break down the wood structure by separating the fibres. This method is called the thermo-mechanical process.

- **tolerance**
  1. the ability of an organism or biological process to subsist under a given set of environmental conditions. The range of these under which it can subsist, representing its limits of tolerance, is termed its ecological amplitude. For trees, the tolerance of most practical importance is their ability to grow satisfactorily in the shade of and in competition with other trees.
  2. a characteristic of trees that describes the relative ability to thrive with respect to the growth factors (light, heat, water nutrients, anchorage). For instance a 'shade tolerant' species may thrive at low light levels.

- **tolerant species**
  a species of tree that has the ability to grow in the shade of other trees and in competition with them.

- **tolerant trees**
  trees that can grow in the shade of other trees.

- **tongs**
  a pair of curved arms with sharp points, pivoted like scissors to bite into a log for yarding or loading.

- **top**
  to cut off the top of a tree down to a utilisable diameter for use as a spar. (Somewhat rare these days, as mobile, portable spars and towers are the rule.)

- **top height**
  the average height of the hundred trees of largest diameter per hectare.

- **topographic break**
  a distinct change in the slope of the land.

- **topography**
  the physical features of a geographic area, such as those represented on a map, taken collectively; especially, the relief and contours of the land.

- **total chance planning**
  early planning over an entire development area for the best overall realisation of all objectives identified by broader planning.

- **total dissolved solids**
  a measure of total filterable residue in water. Expressed as mg/l.

- **total fertility rate**
  births expected per woman during her reproductive years (15-49).

- **total logging cost**
  the sum of road and yarding/skidding costs per MBF.

- **total resource plan**
  a plan for long-term forest management over an entire area, such as a watershed. The plan identifies known resource values, capabilities and sensitivities; confirms or refines management objectives for those values; and establishes detailed management guidelines by which to achieve those objectives on the ground.

- **tower**
  a steel mast or framework, generally protable, used in-

stead of a spar tree for cable yarding.
Standing tower: A vertical, telescoping steel tower, usually cylindrical in cross-section, which is portable and transported fromlanding to landing for cable yarding.
Swing tower: A leaning tower, usually of lattice-box steel construction, mounted on a turntable and capable of landing logs to either side as well as in front of the yarder.

- **TPD**
  Tons Per Day.

- **trade-off**
  a management decision whereby there is a reduction of one forest use in favour of another, such as a reduced timber yield in favour of improved wildlife habitat. In some cases, a management decision favouring one use in one location, is offset by a reverse decision favouring another use in another location.

- **traditional biotechnology**
  specifically within forestry, this involves the selection of trees with desirable traits and breeding them to produce beneficial offspring.

- **traffic lanes**
  numeric code indicating the number of traffic lanes contained within the travelled way of an existing road management section.

- **trait**
  a distinguishable characteristic of an organism.

- **transformation**
  a transfer of foreign DNA into the cell of an organism to change its genetic makeup. This is a natural process for many bacteria.

- **transgenic plant**
  Genetically modified plant expressing traits not found in its wild relatives. Many are more tolerant of herbicides, are resistant to insect or viral pests, or express modified versions of fruit or flowers.

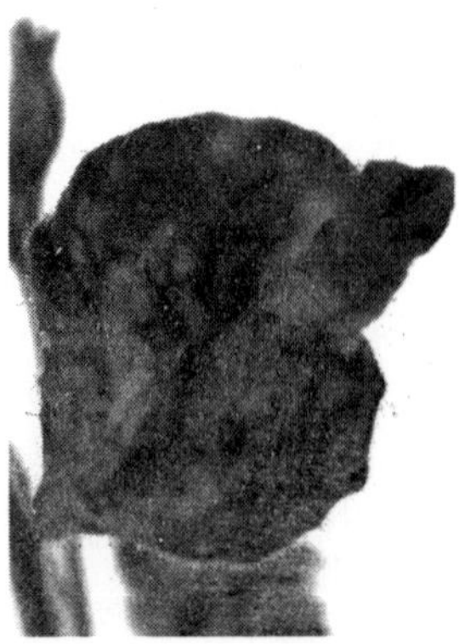

- **transgenic**
genetically engineered to contain DNA from an external source.

- **transpiration**
the process by which plants absorb water from their roots, move it up through the plant (via the xylem), pass it through pores (stomata) in the leaves and other plant parts and then evaporate it into the atmosphere as water vapour.

- **transplanting**
moving seedlings from one place to another

- **treatment prescription**
operational details required for carrying out individual silviculture activities such as site preparation and planting.

- **treatment season**
the season or year the planned treatment activity will be carried out.

- **treatment unit**
the geographic unit of productive forest land area designated in a prescription for a specific silviculture activity or series of treatments.

- **tree**
a woody plant usually having one or more perennial stem at least 3 inches d.b.h. at maturity, a more or less definitely formed crown of foliage and a height of at least 16 feet at maturity.

- **tree age**
the total age of the above ground stem of the tree in years (not the age of the rootstock or the total age from seed). Total age is usually the annual ring count to the pith of the tree at breast height plus an estimate of the number of years it took the tree to reach breast height. This must be an estimate based on local knowledge.

- **tree class**
the overall quality of live trees 1.0 inches d.b.h. and larger.
Growing stock: Merchantable tree.
Rough cull tree: Tree that does not meet regional merchantability standards be-

cause of excessive sound cull. May include noncommercial tree species.
Rotten cull tree: Tree that does not meet regional merchantability standards because of excessive unsound cull. May include noncommercial tree species.

- **tree farm**

1. a privately owned forest or woodland in which timber crop production is a major management goal.
2. a privately owned woodland where sustainable forest management is the primary objective.

- **tree history**

a classification of the current status (living or dead) of a tally tree.
Live trees: Survivors-At initial measurement any live tree on a plot. At remeasurement: survivor-tree previously measured and still living.
Ingrowth: At remeasurement. Ingrowth- newly established seedling or a tree that passed the minimum size for measurement since the previous measurement. Ongrowth: At remeasurement:- a tree not previously measured or previously measured on a smaller plot or with some different sampling probability which has by the time of remeasurement grown enough to be measured on a larger plot or with a different sampling probability.
Dead trees: Cut Trees: At initial measurement- A tree cut within the last n years. The number of years, n, within which cut trees are counted will be determined by field units. At remeasurement: a tree present and alive at previous measurement but since cut. Nonsalvable mortality: At initial measurement: A tree found dead and judged to have died within the past n years. The number of years, n, will be determined by Regions. At remeasurement: a tree of any size which was alive at the previous measurement, but has since died. Salvable mortality: At initial measurement: A tree found dead, judged to have died within the past n years and which, at the time of inventory met the growing stock standards for mer-

chantability. At remeasurement: a tree alive at previous measurement and has since died; at time of death it would meet the growing-stock standards for merchantability.

- **tree length (height)**
  the total span of a tree from ground level along bole to tip of tree. Entire tree, excluding the unmerchantable top and limbs.

- **tree spacing**
  the distance between trees, which is most often regulated at the time of planting or during a harvest or thinning operation. Spacing, like stand density, affects understory vegetation, seed production, growth rate and wildlife habitat.

- **tree top condition**
  an indication as to whether or not the top of the tree is intact. The various tree top conditions include: a) Intact live top, b) Broken top, c) Intact dead top. Because tops may be either dead or merely defoliated, we can use this category with caution.

- **tree volume**
  the amount of wood in a tree. This may be expressed in board feet or cubic feet. It may be gross volume or net volume (gross less defects).

- **tree-length harvesting system**
  a method of harvesting that includes felling a tree, cutting of the top and delimbing it before transport to a mill.

- **tree-length logging**
  felling and transporting the trimmed bole in one piece, whenever possible, for crosscutting at a landing or mill. Unbucked, limbed and topped trees.

- **tree-length system**
  a harvesting system in which the tree is felled and delimbed in the woods, and hauled to the landing (usually by a skidder).

- **trichocyst**
  a type of extrusome which, when extruded, takes the form of a fine stiff filament. Typical of Paramecium, but used inadvertently to refer to other types of extrusomes.

- **trim allowance**
  the extra length allowed when bucking logs to account for end injury, uneven cut and trimming of boards in the mill. Trim allowance may vary by local and scaling rule used, as well as by local tradition. In California, for example, this is commonly six inches of trim for a 16 foot or less log and 12 inches for a 32 foot log.

- **trophic**
  said of organisms which are active and feeding and contrasts with the encysted state, theronts or swarmers. May also be used to refer to those aspects of metabolism associated with growth.

- **trunk**
  the stem or body of the tree which supports the crown.

- **tuft**
  bunching of twigs or needles.

- **turbidity**
  sediment suspended in the water column.

- **turn**
  the logs brought to the landing in any one yarding or skidding cycle.

- **turnout**
  a widening in the roadway where a vehicle may pull or park to allow other vehicles to pass safely.

- **turpentine**
  a paper making by-product commonly used as a paint thinner and solvent.

- **twister**
  a line between anchors used for support by twisting the line with a sturdy stick to reduce its length.

- **tylosis**
  outgrowth of a cell membrane from a ray or axial parenchyma cell through a pit in a xylem vessel wall into the vessel, partially or completely blocking the lumen of the vessel.

- **ultimate strength**
  see **breaking strength**.

- **unclamped carriage**
  a carriage which holds position without the use of a skyline clamp. See **Clamped Carriage**.

- **undercut**
  the wedge-shaped first cut made when felling a tree which determines the direction of fall.

■ **underplanting**
planting young trees under the canopy of an existing stand.

■ **understorey**
1. any plants growing under the canopy formed by other plants, particularly herbaceous and shrub vegetation under a tree canopy.
2. the layer formed by the crowns of smaller trees in a forest. The trees beneath the forest canopy.-Foliage layer beneath the forest canopy.
3. young trees that are growing beneath the tall mature trees in a timber stand.
4. the smaller trees beneath the forest canopy.
5. the smaller vegetation (shrubs, seedlings, saplings, small trees) within a forest stand, occupying the vertical zone between the overstory and the herbaceous plants of the forest floor.

■ **understory fire**
a fire burning in the understory, more intense than a surface fire and with flame lengths of 1-3 meters.

■ **undulipodium**
a term preferred by some workers, usually in the US, for eukaryotic cilia and flagella.

■ **uneven-aged management**
silvicultural system in which individual trees originate at different times and result in a forest with trees of all ages and sizes. Harvest cuts are on an individual-tree selection basis.

■ **uneven-aged silvicultural system**
a silvicultural system designed to create or maintain and regenerate an uneven-aged stand structure. Single-tree and group selection are uneven-aged silvicultural systems.

■ **uneven-aged stand**
a stand of trees containing three or more age classes. In a balanced uneven-aged stand, each age class is represented by approximately equal areas, providing a balanced distribution of diameter classes.

- **uneven-aged stand management**
  the practice of managing a forest for various age classes by periodically selecting and harvesting individual trees of groups of trees.
- **uneven-aged trees**
  a group of trees that have a wide range of ages.
- **unhook**
  to remove chokers from logs at the landing.
- **unicellular**
  consisting of but a single cell.
- **unit**
  see **cutting unit**.
- **unloaded deflection**
  deflection when no logs are being supported on the skyline.
- **unmanaged forest land**
  forest land that is not subject to management under a forest management plan.

- **unmerchantable**
  of a tree or stand that has not attained sufficient size, quality and/or volume to make it suitable for harvesting.
- **unmerchantable wood**
  material that is unsuitable for conversion to industrial wood products due to size, form, or quality. May include rough, rotten, and dead trees; the tops, limbs and cull sections from harvested trees; or small and noncommercial trees.
- **unrecovered timber**
  timber as described in the Provincial Logging Residue and Waste Management Procedures Manual.
- **unrecovered volume**
  timber that is within the cutting specifications of the minimum utilisation standards of the cutting authority and not removed from the area.
- **unsalvaged losses**
  the volume of timber destroyed by natural causes such as fire, insect, disease or blowdown and not harvested, including the timber actually killed plus any re-

sidual volume rendered non-merchantable.

- **unstable or potentially unstable terrain**
  an area where there is a moderate to high likelihood of landslides.

- **upgrade**
  to increase in value, or improve

- **uplands**
  terrain not affected by water table or surface water or else affected only for short periods so that riparian (hydrophilic) vegetation or aquatic processes do not persist.

- **upper stem portion**
  sawtimber tree bole extending from above the merchantable top to a minimum 4-inch top diameter outside bark or to the point where the central stems break into limbs.

- **urban and suburban areas**
  areas within the legal boundaries of cities and towns; suburban areas developed for residential, industrial, or recreational purposes; school yards; cemeteries; or other nonforest land not included in any other specified land use class.

- **urban forestry**
  the cultivation and management of trees and forests for their present and potential contributions to the physiological, sociological and economic well-being of urban society.

- **urban forestry**
  the professional management of natural resources in and around urban areas, including trees and associated vegetation, wildlife, and open space.

- **urban wood**
  used pallets, wooden shipping crates and clean construction wood diverted from the waste stream and chipped for use in making particleboard and MDF.

- **uredial stage (uredium)**
  a stage of the rust fungi; a fruiting structure usually appearing as a pustule bearing the repeating spore stage that leads to disease increase.

- **utility log**
  a log not meeting the grade requirements of a peeler or saw log, but which will pro-

duce at least 50% of its gross volume in pulp chips. AKA 'chip log'.

- **utilisation (of forage and browse)**
  the level of forage and browse use on a site. For herbaceous species, it is measured as a percentage of the current year's growth removed; for browse species, it is measured as a percentage of stem ends removed.

- **utilisation standards**
  the dimensions (stump height, top diameter, base diameter, and length) and quality of trees that must be cut and removed from Crown land during harvesting operations.

- **vacuole**
  a structure in a cell enclosed by a membrane. Usually food vacuoles (associated with the digestion of food) or contractile vacuoles (association with the excretion

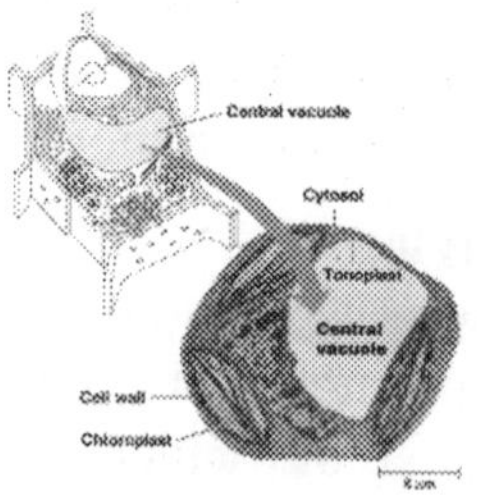

of fluid). Small vacuoles may be called vesicles.

- **values-at-risk**
  the specific or collective set of natural resources and man-made improvements/developments that have measurable or intrinsic worth and that could or may be destroyed or otherwise altered by fire in any given area.

- **variable area plot sampling method**
  a method of timber cruising commonly used for industrial timber cruising in which sampling area (plot size) varies with tree diameter.

- **variable cost**
  the total cost of operating a machine ($/unit) attributable to actual work, such as fuel, track wear, etc.

- **variable retention**
  a relatively new silvicultural system that follows nature's model by always retaining part of the forest after harvesting. Standing trees are left in a dispersed or aggregated form to meet objectives such as retaining old growth struc-

ture, habitat protection and visual quality. Variable retention retains structural features (snags, large woody debris, live trees of varying sizes and canopy levels) as habitat for a host of forest organisms. There are two types of variable retention:

Dispersed retention: retains individual trees scattered throughout a cutblock,

Aggregate (group) retention: retains trees in clumps or clusters.

The main objectives of variable retention are to retain the natural range of stand and forest structure and forest functions. With retention systems, forest areas to be retained are determined before deciding which areas will be cut. This system offers a range of retention levels. The system also provides for permanent retention of trees and other structures after regeneration is established. Variable retention can be implemented with a range of harvesting systems and can be combined with traditional silvicultural systems such as shelterwood or selection.

- **vector**

any DNA containing structure that is used to transfer DNA into an organism. The most commonly used vectors are plasmid DNA, bacteriophages or viruses.

- **vegetation density**

number of individual plants of a given species in a unit of area. The relative density of a species is the percentage the number of its individuals is of the total number of individuals of all species in the sample.

- **vegetation height**

the vertical distance from ground level to the top of an individual plant or canopy.

- **vegetative lot**

a quantity of vegetative material or vegetative propagules having the same species, source and year of collection.

- **vegetative material**

plant parts or tissues used to produce vegetative propagules through asexual means.

- **vegetative propagation**
an asexual reproduction process of trimming growing stems from aplant and allowing them to root.

- **vegetative propagules**
plants produced through asexual means.

- **vehicle side-tracking**
the lateral displacement of vehicles on a curve caused by the length of the vehicle manoeuvring through the turn; the wider path that the rear of a vehicle takes when negotiating a curve.

- **ventilation index**
a term commonly used in air pollution meteorology. The VI is a numerical value relating to the potential of the atmosphere to disperse airborne pollutants from a stationary source (such as smoke from a prescribed fire). It is calculated by multiplying the mixing height by the average wind speed in the mixed layer.

- **vernal or autumnal ponds**
a class of wetland characterised by small, shallow, temporary pools of fresh water present in spring and fall, which typically do not support fish but are very important breeding grounds for many species of amphibians. Some species are totally dependent upon such ponds; examples are spring peepers and mole salamanders.

- **vertical integration**
in the forest products industry, a vertically integrated company grows its own trees, makes products from them, then makes other products from fiber leftovers from the initial manufacturing operation, then converts and adds value to all these products.

- **vertical structure**
the arrangement of plants in a given community from the

ground (herbaceous and woody shrubs) into the main forest canopy; a complex vertical structure is characterised by lush undergrowth and successive layers of woody vegetation extending into the crowns of dominant and codominant trees.

- **veteran**
  in growth and yield, a tree that is at least 30 years older than the age of the main stand. In multi-layered or complex-layered stands, a tree that is at least 100 years older than the oldest sample tree of the main stand.

- **viewshed**
  a physiographic area composed of land, water, biotic, and cultural elements which may be viewed and mapped from one or more viewpoints and which has inherent scenic qualities and/or aesthetic values as determined by those who view it.

- **virgin forest**
  a forest that has never been harvested or altered by humans.

- **virgin timber**
  timber from an original forest that has not been previously disturbed or influenced by human activity. Many consider 150 year old trees virgin.

- **visibility sensitivity**
  the determination of how rapidly visibility can be reduced.

- **vision statement**
  an explicit statement of what you want to do, be or have in the future. It must be positive and detailed enough for people to form a mental picture. It must be inspiring, challenging and worth the effort.

- **visual absorption capability**
  a component of the visual landscape inventory that rates the relative capacity of a landscape to absorb visual alterations and still maintain its visual integrity.

- **visual green-up**
  see **green-up**.

- **visual impact assessment**
  an evaluation of the visual impact of resource development proposals on forest landscape.

- **visual landscape analysis**
  the process of recommending visual quality objectives

based on the visual landscape inventory, number of viewers, level of concern and in consideration of other values.

- **visual landscape inventory**
  the identification, classification, and recording of the location and quality of visual resources and values.

- **visual landscape management**
  the identification, assessment, design and manipulation of the visual features or values of a landscape, and the consideration of these values in the integrated management of provincial forest and range lands.

- **visual landscape unit**
  a landform or portion of a landform visible from one or more viewpoints identifying relatively homogeneous visual landscape features.

- **visual quality**
  the character, condition, and quality of a scenic landscape or other visual resource and how it is perceived, preferred, or otherwise valued by the public.

- **visual quality**
  degree of obstruction or contrast degradation of viewing scene due to air containments or weather.

- **visual range**
  the distance at which a large (half a degree) black or dark object disappears from view.

- **visual resource management class**
  an assessment of the relative visual resource quality on National Forest system lands as it relates to potential resource use and or development. The five visual quality objectives, each representing a different degree of acceptable alteration of the natural

- **visual sensitivity**
  a component of the visual landscape inventory that estimates the sensitivity of the landscape based on the visual prominence or importance of features, conditions that affect visual perception, and social factors that contribute to viewer perceptions.

- **visually sensitive areas**
  viewsheds that are visible from communities, public

use areas, and travel corridors, including roadways and waterways, and any other viewpoint so identified through referral or planning processes.

- **vital events**
  births and deaths.

- **volume table**
  a table showing the estimated average tree or stand volume based on given tree measurements, usually diameter and height.

- **vulnerable species**
  see **sensitive/vulnerable species**.

- **WTC**
  Whole Tree Chips.

- **waste**
  the volume of timber left on the harvested area that should have been removed in accordance with the minimum utilisation standards in the cutting authority. It forms part of the allowable annual cut for cut-control purposes.

- **waste area**
  a pre-approved site for disposal of excavations.

- **water**
  a compound consisting of two parts hydrogen and one part oxygen, H_2O, or Hydrogen Hydroxide. Pure water is a colourless, odourless, transparent liquid.

- **water areas**
  areas within a land mass persistently covered by water. Examples are bays, estuaries, streams, reservoirs and lakes.

- **Census Water**
  streams, sloughs, estuaries, and canals more than 1/8 of a statute mile in width. Lakes, reservoirs and ponds more than 40 acres in size.

- **Non-Census Water**
  streams, sloughs, estuaries, and canals more than 120 feet and less than 1/8 of a mile wide. Lakes, reservoirs and ponds 1 to 40 acres in size.

- **water bar**
  1. a diagonal ditch or hump in a trail that diverts surface

water runoff to minimise soil erosion.
2. a mound of soil built across a light-duty road, skid trail or trail, or fireline for the purpose of diverting surface water onto porous forest soil.
3. a shallow trench and/or hump across a trail or road (generally tied into the uphill side of the slope) for the purpose of carrying water runoff into the vegetation, duff, ditch or dispersion area. Water bars are used to minimise erosion by diverting the waters potential velocity and volume.

- **water bomber**
  see **airtanker**.

- **water control**
  management of water (both surface and subsurface) to maintain plant growth, water quality, wildlife habitat and fire control.

- **water course**
  a channel where water flows either perennially or intermittently. Can be also used to include bodies of open water.

- **water flow velocity**
  the average velocity of water flowing through a cross-section of a stream.

- **water management**
  the planned development, distribution and use of water resources.

- **water marketing**
  the idea that farmers and other users with water rights that save through conservation or growing less water-thirsty crops should be able to sell or lease the water they save to industries and cities rather than losing their rights to the water.

- **water pollution**
  any introduction of foreign material into water or other impingement upon water which produces undesirable changes in the physical, biological, or chemical characteristic of that water.

- **water quality**
  the physical, chemical and biological properties of water.

- **water repellency**
  the resistance to soil wettability, which can be increased by intense fires.

- **water resources**
  the supply of water in a given area or basin inter-

preted in terms of availability of surface and underground water.

- **water uses**

the status of water uses subject to State water laws.

Domestic (Household): Includes household water, watering of lawns, gardens, etc., for Forest Service administrative sites or other residential areas not served by a municipal water system, and not directly related to recreational use. For private residences, includes watering of lawns and gardens up to 1/2 acre.

Domestic (Recreational): Includes camp and picnic grounds, special use concessionaires, recreational residences, lodges, restaurants, etc., not served by a municipal water system. This code should be used for all water systems developed solely to serve recreational purposes regardless of the number of persons served or the number of service connections.

Fish and Wildlife: Includes, but is not limited to, storage of water that is either retained in the reservoir or released downstream to support these purposes. Also includes instream reservation of natural flow for these purposes.

Flood Control: Water (stored, directed or rerouted) for the sole purpose of flood control or abatement.

Industrial: Water used in manufacture or production.

Instream: In-channel flow for purposes other than those under Recreation, Fish and Wildlife, and Navigation, and when there are multiple instream purposes.

Irrigation: Application of water for the production of crops or the maintenance of large (more than 1/2 acre) areas of lawns, gardens, etc. This includes Forest Service pastures and golf courses.

Mining: Any application of water to mining processes (e.g. hydraulic), for drilling and on concentrator tables.

Municipal: Water provided for human consumption and other purposes through a city, town, community, water district or other formalised water system regularly serving at least 25 individuals at least 60 day's out of the year or provides at least 15 service connections.

Navigation: Use of water specifically for the maintenance of navigation for commerce or trade. Excludes water used for recreational boat travel.

Power: Water used for hydroelectric and hydromechanical power generation.

Recreation: Any use other than above including impoundment or instream flow specifically intended for recreational use. This also includes water required for Wild and Scenic Rivers.

Stockwater: Use for domestic livestock grazing on National Forest, other public or private lands.

- **water width**

the width of the top of the water prism from channel bank to channel bank. The line of measure is perpendicular to the flow direction.

- **waterbar**

a shallow ditch excavated across a road at an angle to prevent excess surface flow down the road surface and subsequent erosion of road surface materials. A small excavation across a road to collect and divert roadway surface water flow.

- **watercourse**

a natural stream or source or supply of water, whether usually containing water or not, such as a lake, river, creek, spring, ravine swamp and gulch.

- **waters of the state**

any surface or underground waters, except those surface waters which are not confined but are spread and diffused over the land. This includes all lakes, ponds, marshes, rivers, streams, ditches, springs and water from underground aquifers, regardless of their size or location.

- **watershed**

1. an area of land that collects and discharges water into a single main stream through a series of smaller tributaries.

2. a sloping area of land that collects, directs, controls, and discharges the flow of rainwater into a single main stream through a series of smaller tributaries, or all

land and water within the confines of a drainage basin.
3. the common land area that drains into a lake, river or river system.
4. a region or area defined by patterns of stream drainage. A watershed includes all the land from which a particular stream or river is supplied.

- **watershed assessment**

evaluates the present state of watersheds and the cumulative impact of proposed development on peak flows, suspended sediment, bedload and stream channel stability within the watershed.

- **watershed integrity**

1. refers to a stable overall physical condition of the watershed (bedrock, landforms, soils, drainage ways) within which transfers of energy, matter and, especially of water occur. It is prerequisite for the security of forest and stream ecosystems.
2. the planned use of drainage basins in accordance with predetermined objectives.

- **water-soaked**

a dull green coloration of diseased tissues due to membrane leaking of cellular contents into intercellular spaces.

- **wedging**

the use of plastic wedges driven into the backcut to direct the tree into its fall and to avoid setbacks on to the chainsaw.

- **weeding**

a release treatment in stands during the seedling stage that eliminates of suppresses undesirable vegetation regardless of crown position.

- **wet line**

a wet line is a line of water, or a water and chemical fire retardant, sprayed along the ground, to serve as a temporary fireline which will stop a low-intensity fire.

- **wetland**

1. a swamp, marsh or other similar area that supports natural vegetation that is distinct from adjacent upland areas.
2. those areas that are inundated by surface or ground water with a frequency sufficient to support, and under normal circumstances do or would support, a prevalance

of vegetation or aquatic life that requires saturated soil conditions for growth and reproduction. Wetlands generally include swamps, marshes, bogs, and similar areas such as sloughs, potholes, wet meadows, river overflows, mud flats and natural ponds.

3. those areas that are inundated by surface or ground water with a frequency sufficient to support, and under normal circumstances do or would support, a prevalence of vegetation or aquatic life that requires saturated soil conditions for growth and reproduction. Wetlands generally include swamps, marshes, bogs, and similar areas such as sloughs, potholes, wet meadows, river overflows, mud flats and natural ponds.

4. geographic areas characteristically supporting hydrophytes, hydric soils, and some saturation or flooding during the growing season.

5. an area where water is near, at or above the land surface long enough to be capable of supporting aquatic or hydrophytic (water loving) vegetation and/or which has soils indicative of wet conditions.

6. areas which are either transitional between land and water (where the water table is at or near the land surface) or areas of land which are covered by shallow water (such as marshes, swamps, bogs, and fens). Wetlands fulfill an essential role in our landscapes by maintaining water quality, stabilising shores and stream banks, controlling floods and erosion and providing critical habitat to many plant and animal species.

- **whistle punk**

a somewhat archaic term for the person who controls the whistle on a cable yarding operation. Back in the old days, before radio signals, the whistle punk was a member of the yarding crew who was stationed where he could see the entire operation and could control the whistle signals through the use of a pull line back to the yarder.

■ **whole tree**

1. refers to felled trees, unbucked with limbs and tops intact.

2. all components of a tree, except the stump. Also known as a full tree.

■ **whole-tree chips**

chips made from a whole-tree.

■ **widow maker**

a loose limb, top, piece of bark or anything loose in a tree that may fall on a logger.

■ **wilderness**

an area of land generally greater than 1000 ha that predominantly retains its natural character and on which the impact of man is transitory and, in the long run, substantially unnoticeable.

■ **wilderness area**

1. a part of the provincial forest designated by order in council as a wilderness area.

2. an area established by the federal government to be managed and preserved in an essentially untouched condition.

■ **wildfall**

trees or parts of trees felled by high winds.

■ **wildfire**

1. an unplanned or unwanted natural or human-caused fire, or a prescribed fire that threatens to escape its bounds.

2. unplanned fire requiring suppression action. Can be contrasted with a prescribed fire that burns within prepared lines, enclosing a designated area, under predetermined conditions.

3. a fire, naturally caused or caused by humans, that is not meeting land management objectives and thus requires suppression.

4. uncontrolled fires occurring in forestland, woodland, brushland and grassland.

■ **wildland urban interface**

a popular term used to describe an area where various structures (most notably private homes) and other human developments meet or are intermingled with forest and other vegetative fuel types.

- **wildlife**

  1. raptors, threatened species, endangered species, game and other species of vertebrates prescribed as wildlife by regulation.
  2. a broad term that includes nondomesticated vertebrates, especially mammals, birds and fish.

- **wildlife & fish habitat capability**

  the ability of a specified area to support a species expressed in terms of numbers of animals or habitat capability index.

- **Wildlife & Fish User Days (WFUD)**

  a 12-hour day in which a person participates in a wildlife- or fish-related recreation activity.

- **wildlife habitat**

  areas of land and water that support specific wildlife or groups of wildlife.

- **wildlife management**

  the application of scientific and technical principles to wildlife populations and habitats to maintain such populations (particularly mammals, birds and fish) essentially for recreational and/or scientific purposes.

- **wildlife tree**

  a live tree retained to become future snag habitat.

- **wildling**

  a seedling naturally reproduced outside of a nursery, used in reforestation.

- **wilt**

  a common symptom of disease due to a loss of turgor and resulting in subsequent drooping and collapse of the foliage or succulent tissues.

- **winch**

  1. steel spool connected to a power source. Used for reeling or unreeling cable. Also known as drum.
  2. a powered drum used to reel in or pay out cable for yarding or hoisting.

- **wind shake**

  crack or separation in a tree's wood structure caused by high winds.

- **wind speed**

  the wind speed at the height of interest (surface, plume height or upper air).

- **windfall**

  tree or trees that have been uprooted or broken off by the wind. Also known as blow down.

- **windfirm**
  tree(s) able to withstand strong winds.

- **windrow**
  1. an accumulation of slash, branchwood and debris on a harvested cutblock created to clear the ground for regeneration. Also refers to an accumulation of fill or surfacing material left on the road shoulder as a result of grading operations.
  2. a long, narrow row of vegetation, debris, and some soil created during site preparation and clearing operations.
  3. brushwood, slash, etc., concentrated (usually by machine) along a line, to clear the intervening ground for regeneration.
  4. slash, logs, or other material piled in a more or less continuous line to clear the intervening ground.

- **windthrow**
  trees uprooted by excessive wind. Shallow-rooted trees are almost always affected. See **blowdown**.

- **winter range**
  a range, usually at lower elevation, used by migratory deer, elk, caribou, moose, etc., during the winter months and typically better defined and smaller than summer range.

- **wire rope**
  a flexible steel rope made up of numerous wire strands twisted hellically together about a wire or fibre core.

- **witches' broom**
  a massed proliferation of the branches of a woody plant.

- **withdrawals**
  lands that have been removed or segregated from the operation of some or all of the public land laws through Executive or Congressional action.

- **witness tree**
  tree used by surveyors to mark the location of a survey corner; the tree is located near the survey corner and is inscribed with survey data. Also known as a bearing tree.

- **wolf tree**
  1. a dominant tree, which is often a remnant from a previous stand, having a broad crown and many limbs.

2. large rough tree, generally not good for lumber.
3. a large, excessively branchy tree which occupies more space in the forest than surrounding trees. Wolf trees have high wildlife and aesthetic value, but little if any timber value.

- **wood chipper**
a machine that reduces logs to wood chips.

- **wood conversion**
transformation of natural timber into any kind of commercial product. Includes all activities from commercial timber (log) delivery to the log yard at the initial commercial processing facility to the final product form offered for commercial sale as a consumer product.

- **wood fibre**
the most important ingredient in pulp and paper; it is composed mainly of cellulose.

- **wood lot**
a piece of land on which trees are cultivated.

- **wood pulp**
fibre from wood with varying degrees of purification that is used for the production of paper, paper board and chemical products.

- **wood veneers**
a thin surface layer of fine wood laid over a base of common material.

- **wood/forest**
generally speaking, a forest is a large wood or a collection of woods although the two terms are often interchangeable.

- **woodfree**
a paper that is produced from pulp that has undergone chemical rather than mechanical breakdown and is free of the chips found in groundwood and mechanical pulp.

- **woodland floor**
all dead vegetative matter on the mineral soil surface in the woodland, including litter and unincorporated humus.

- **woodlot**
the wooded portion of a private property upon which small-scale forestry operations are carried out.

- **working plan**
see **management** and **working plans**.

- **working strength**
  the maximum allowable load a wire rope or other material should be stressed to and still maintain the desired 'factor of safety'. Usually this load is less than the elasticity limit of the wire rope or other material.

- **wrapper**
  see **binder**.

- **xenosome**
  a foreign body. Used to refer both to bits of debris which may be incorporated into the shells of some testate amoebae, and also to symbiotic algae lying inside other cells. Probably best restricted to the first use.

- **xylem**
  1. the principal strengthening and water-conducting woody tissue of stems, leaves and roots.
  2. the main part of the tree trunk made up of tiny tubes which transport water and minerals from the roots up the trunk and branches to the leaves.

- **yarder**
  a machine or system of winches used to haul logs into a landing. Often combined with a portable tower. Technically the yarder is just the power and winch system, but in practicality today, it is the whole machine including the tower.

- **yarder engineer**
  the person who operates and is responsible for the yarder.

- **yarding**
  1. the act or process of conveying logs to a landing. In common practice, yarding is often reserved for cable logging, while skidding is used for ground based logging.
  2. in logging, the hauling of felled timber to the landing or temporary storage site from where trucks (usually) transport it to the mill site. Yarding methods include cable yarding, ground skidding and aerial methods such as helicopter and balloon yarding.

- **yield analysis**
  the study of forest yield over time using mathematical models and inventory data.

- **yield curve**
  a representation of stand volume, usually as a func-

tion of stand age, in graphical or tabular form.

■ **zero sum**
a game or scenario where one wins all and the other loses all. Often used to describe the economy against the environment. Sustainable development rejects this zero sum scenario. It's win/win rather than win/lose.

■ **zone line**
narrow brown or black lines in decayed or decaying wood.

■ **zooplankton**
small aquatic invertebrates that live in the sunlit waters of streams, lakes, or oceans and feed on algae and other invertebrate animals. Zooplankton are the most prolific life forms on earth.

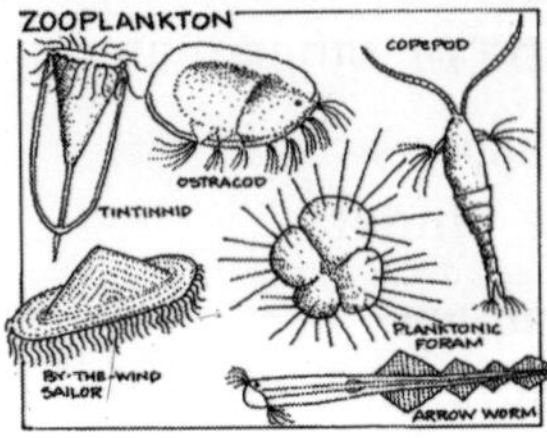